Horticultural Crops Breeding

Horticultural Crops Breeding

Edited by:

Kiran Raj and Amit Kanawjia

Delve PUBLISHING

www.delvepublishing.com

Horticultural Crops Breeding

Kiran Raj and Amit Kanawjia

Delve Publishing
224 Shoreacres Road
Burlington, ON L7L 2H2
Canada
www.delvepublishing.com

Email: orders@arclereducation.com

ISBN: 978-1-77407-789-4 (Hardcover)

ABOUT THE EDITORS

Dr. Kiran received her education at the universities in Bangalore (B.Sc. Biotechnology), Allahabad (M.Sc Ag.) and CSKHPKV, Palampur (Ph.D Plant Breeding and Genetics.) She has throughout a brilliant academic record. She worked at G.B. Pant University of Agriculture and Technology, Pantnagar and GKU,Bathinda . She has a number of books ,research papers to her credit. Her research interests range from Plant breeding & Genetics, Seed science & technology, Biotechnology, Bioinformatics and Plant tissue culture .She has also voluntarily worked as a coordinator, society for social development ,as a community development trainer, Raipur and as a public support executive, Navjeeven foundation, Kangra HP, India

Dr. Amit Kanawjia, did his B.Sc. (Ag.) from AAIDU, Allahabad and M.Sc. (Ag.) Horticlture for IAS, BHU, Varanasi. Dr. Kanawjia completed his Ph. D. (Horticulture) from SHIATS, Allahabad. He has more than 6 years experience in teaching and research. He has supervised 15 students for M.Sc. (Ag.) Horticulture dissertations in the area os fruits, vegetables and flowers as a Chairman and member of advisory committee. Besides, he qualified JRF conducted by ICAR, New Delhi for M.Sc. (Ag.). He organized National Conference and Workshops, as an Organizing Secretary. He has published more than 30 research papers, articles and book chapters in Journals and Magazines of International repute. He is life member of Indian Society of Vegetable Sciences and Organic Agriculture Society of India.Presently Dr Kanawjia is working as Subject Matter Specialist (Horticulture) in Krishi Vigyan Kendra, Jalaun administered by Banda University of Agriculture and Technology, Banda, India.

TABLE OF CONTENTS

GLOSSARY OF TERMS

A

Abandoned - having been deserted or left

Aboriginal - inhabiting or existing in a land from the earliest times or from before the arrival of colonists; indigenous

Accessions – the attainment or acquisition of a position of rank or power

Aesthetically – in a way that gives pleasure through beauty

Agrobacterium - a genus of Gram-negative bacteria established by H. J. Conn that uses horizontal gene transfer to cause tumors in plants

Agrobacterium tumefaciens - the causal agent of crown gall disease in over 140 species of eudicots. It is a rod-shaped, Gram-negative soil bacterium

Agronomists - an expert in the science of soil management and crop production

Agronomy - Agronomy is the science and technology of producing and using plants in agriculture for food, fuel, fiber, and land restoration

Allopolyploid - having more than two haploid sets of chromosomes that are dissimilar and derived from different species

Amenable – open and responsive to suggestion

Anaphase - the third stage of cell division, between metaphase and telophase, during which the chromosomes move away from one another to opposite poles of the spindle

Androdioecy - a sexual polymorphism in which populations are composed of co-sexual and male plants. Very rare in flowering plants.

Angiosperm - a plant of a large group that comprises those that have flowers and produce seeds enclosed within a carpel, including herbaceous plants, shrubs, grasses, and most trees

B

Backcrossing - cross (a hybrid) with one of its parents or an organism with the same genetic characteristics as one of the parents

Bibliometric – is the use of statistical methods to analyze books, articles, and other publications.

Budding - graft a bud of (a plant) on to another plant

C

Cereals - a grain used for food, for example wheat, maize, or rye

Chutneys - a spicy condiment of Indian origin, made of fruits or vegetables with vinegar, spices, and sugar

Cisgenic - a product designation for a category of genetically engineered plants

Congenial - (of a thing) pleasant or agreeable because suited to one's taste or inclination

Connotes – imply or suggest

Consuming - eat, drink, or ingest (food or drink)

Co-sexuality - condition in which plants reproduce as both maternal and paternal parents as a result of hermaphrodite sex expression. The commonest sexual system in flowering plants.

Cultivars – A cultivar is an assemblage of plants selected for desirable characteristics that are maintained during propagation.

Cytology - the branches of biology and medicine concerned with the structure and function of plant and animal cells.

Cytoplasmic – The jelly like material that makes up much of a cell inside the cell membrane, and, in eukaryotic cells, surrounds the nucleus.

D

Deadheading – remove dead flower heads from

Deficient - not having enough of a specified quality or ingredient

Deleterious – causing harm or damage

Depletion - reduction in the number or quantity of something

Detasseling – is removing the immature pollen-producing bodies, the tassel, from the tops of corn (maize) plants and placing them on the ground.

Devolved - having had power transferred or delegated to a lower level, especially from central government to local or regional administration

Diminishing – cause to seem less impressive or valuable

Diminutive - extremely or unusually small

Dioecious – (of a plant or invertebrate animal) having the male and female reproductive organs in separate individuals.

Disjunction - a lack of correspondence or consistency

Disseminating - spread throughout an organ or the body

Dwindling – gradually diminishing in size, amount, or strength

E

Ecological – relating to or concerned with the relation of living organisms to one another and to their physical surroundings

Edaphic – of, produced by, or influenced by the soil

Elementary - of the most basic kind

Elicit – to get or produce something, especially information or a reaction

Elucidate – to make lucid especially by explanation or analysis *elucidate* a text

Eminently – to a notable degree

Epigenetic – epigenetics is the study of heritable phenotype changes that do not involve alterations in the DNA sequence.

Erucic Acid – a monounsaturated omega-9 fatty acid, which is present in the oil-rich seeds of the Brassicaceae family of plants, particularly rapeseed and mustard

Essence - a property or group of properties of something without which it would not exist or be what it is

F

Floral design - characteristics of individual flowers including their size, structure, sex condition, color, scent, nectar production and degree of herkogamy and dichogamy

Floral display - the number of open flowers on a plant and their arrangement within and among inflorescences. The important functional unit for pollination is usually daily inflorescence size

G

Geitonogamy - mating that results from transfer of pollen between flowers on an individual. Common in mass-flowering species, genetically equivalent to selfing, and causes complete pollen discounting

Genealogy - a line of descent traced continuously from an ancestor

Genetic markers - A genetic marker is a gene or DNA sequence with a known location on a chromosome that can be used to identify individuals or species

Genotypes - the genetic constitution of an individual organism

Gnarled - knobby, rough, and twisted, especially with age

Gossypol – a natural phenol derived from the cotton plant. Gossypol is a phenolic aldehyde that permeates cells and acts as an inhibitor for several dehydrogenase enzymes

Gynodioecy - a sexual polymorphism in which populations are composed of co-sexual and female individuals.

H

Haploid - (of a cell or nucleus) having a single set of unpaired chromosomes

Herbaceous - denoting or relating to herbs (in the botanical sense)

Hereditary – (of a characteristic or disease) determined by genetic factors and therefore able to be passed on from parents to their offspring or descendants

Herkogamy - the spatial separation of anthers and stigmas within flowers. Common in flowering plants and, like dichogamy, generally reduces intraflower self-pollination.

Heterosis – Heterosis, hybrid vigor, or outbreeding enhancement is the improved or increased function of any biological quality in a hybrid offspring

Heterozygous – having two different alleles of a particular gene or genes.

Homozygous - having two identical alleles of a particular gene or genes

Hybrid - bred as a hybrid from different species or varieties

I

Illuminating - light up

Imbalance - lack of proportion or relation between corresponding things

Improvising - produce or make (something) from whatever is available

Inarguably – unquestionably inarguably

Inbreeding - breed from closely related plants ,people, or animals, especially over many generations

Inhabitants – a person or animal that lives in or occupies a place

Inherited – derived genetically from one's parents or ancestors.

Insubstantial - not having physical existence

Intermittent - occurring at irregular intervals; not continuous or steady

K

Kernel - the seed and hard husk of a cereal, especially wheat

Kharif – monsoon crops or autumn crops are domesticated plants like rice that are cultivated and harvested in India, Bangladesh

L

Lodging - Lodging refers to the renting of a short-term dwelling.

M

Meiosis – is a process where a single cell divides twice to produce four cells containing half the original amount of genetic information.

Microspore - the smaller of the two kinds of spore produced by some ferns

Mitosis – a type of cell division that results in two daughter cells each having the same number and kind of chromosomes as the parent nucleus, typical of ordinary tissue growth.

Monoecious – having both the male and female reproductive organs in the same individual

Monoecy - the condition in which plants produce separate female and male flowers. Prevents intraflower self-pollination and enables adjustment of female and male allocation to environmental conditions.

Morphological – relating to the branch of biology that deals with the form of living organisms, and with relationships between their structures

Mutagenesis – is a process by which the genetic information of an organism is changed, resulting in a mutation. It may occur spontaneously in nature, or as a result of exposure to mutagens.

Mutagenic- an agent, such as radiation or a chemical substance, which causes genetic mutation

Mutation – A mutation is a change that occurs in our DNA sequence, either due to mistakes when the DNA is copied or as the result of environmental factors such as UV light and cigarette smoke.

Mythological – relating to, based on, or appearing in myths or mythology

N

Neolithic - The neolithic is the youngest part of the stone Age. The neolithic period is the time when farming was invented and when people started caring for animals, such as cows, sheep, and pigs

Nomadic – living the life of a nomad

Nonexistent - not existing or not real or present

Nurturing - care for and protect (someone or something) while they are growing

Nutrient - a substance that provides nourishment essential for the maintenance of life and for growth

O

Offshoots - a side shoot or branch on a plant

Orchards - a piece of enclosed land planted with fruit trees

P

Parthenocarpy - the development of a fruit without prior fertilization

Perennial -(of a plant) living for several years

Perpetuated – make (something) continue indefinitely

Phenotypic - relating to the observable characteristics of an individual resulting from the interaction of its genotype with the environment

Phenotyping – refers to a quantitative description of the plant's anatomical, ontogenetical, physiological and biochemical properties.

Photosynthate - a sugar or other substance made by photosynthesis

Pistils - the female organs of a flower, comprising the stigma, style, and ovary

Potency - a capacity in embryonic tissue for developing into a particular specialized tissue or organ

Progenitor – a person or thing from which a person, animal, or plant is descended or originates; an ancestor or parent

Progeny - a descendant or the descendants of a person, animal, or plant; offspring

Propagated - breed specimens of (a plant or animal) by natural processes from the parent stock

Propagules – a vegetative structure that can become detached from a plant and give rise to a new plant, e.g. a bud, sucker, or spore

Proponents – a person who advocates a theory, proposal, or course of action

Protandry - The condition in which the male reproductive organs (stamens) of a flower mature before the female ones (carpels), thereby ensuring that self-fertilization does not occur

Protocols - a formal or official record of scientific experimental observations

Pungent - having a sharply strong taste or smell

Pure-breeding - A group of identical individuals that always produce offspring of the same phenotype when intercrossed

Q

Qualitative - relating to, measuring, or measured by the quality of something rather than its quantity

Quantitative - relating to, measuring, or measured by the quantity of something rather than its quality

R

Rabi – rabi harvest are agricultural crops that are sown in winter and harvested in the spring in India and Pakistan

Radical - of, or springing direct from, the root or stem base of a plant

Rapidly - very quickly; at a great rate

Reflectance – the measure of the proportion of light or other radiation striking a surface which is reflected off it.

Regenerate – grow (new tissue) after loss or damage

Replicative - relating to or involving the replication of genetic material or living organisms

S

Sedentary – (of work or a way of life) characterized by much sitting and little physical exercise.

Segregating – set apart from the rest or from each other

Selfing – self-fertilization is the union of male and female gametes and/or nuclei from same haploid, diploid, or polyploid organism

Senescence - biological aging is the gradual deterioration of functional characteristics

Shattering – very shocking or upsetting

Staminate – having stamens but no pistils

Stature – a person's natural height.

Sterile - (of a plant) not able to produce fruit or seeds

Sterility - the quality or condition of being sterile

Stigma - (in a flower) the part of a pistil that receives the pollen during pollination

Strategy - a plan of action designed to achieve a long-term or overall aim

Subtle – (especially of a change or distinction) so delicate or precise as to be difficult to analyze or describe.

Susceptible – likely or liable to be influenced or harmed by a particular thing.

Synergy - the interaction or cooperation of two or more organizations, substances, or other agents to produce a combined effect greater than the sum of their separate effects

T

Testa– the protective outer covering of a seed; the seed coat

Throughput - the amount of material or items passing through a system or process.

Ti plasmid - a tumor inducing plasmid is a plasmid found in pathogenic species of Agrobacterium, including *A. tumefaciens*, *A. rhizogenes*, *A. rubi* and *A. vitis*

Traits - a genetically determined characteristic

Transgenes – A transgene is an exogenous gene that has been introduced into the genome of another organism, and a transgenic species is one whose genome has been genetically altered

Transgenic - relating to or denoting an organism that contains genetic material into which DNA from an unrelated organism has been artificially introduced

Transgenically – *Transgenic* or genetically modified organisms, be they bacteria, viruses, or fungi, serve all kinds of research purposes.

Transgenics - the branch of biology concerned with transgenic organisms

Translocation - the movement of something from one place to another

V

Variability - lack of consistency or fixed pattern; liability to vary or change

Variation - a change or slight difference in condition, amount, or level, typically within certain limits

Vernalization – the cooling of seed during germination in order to accelerate flowering when it is planted.

Vigor - physical strength and good health

Vulnerable - exposed to the possibility of being attacked or harmed, either physically or emotionally

Wicks – absorb or draw off (liquid) by capillary action

Z

Zaid – summer season crops, they grow in long time period mainly from March to June

LIST OF FIGURES

ABBREVIATION

ALS	Amyotrophic Lateral Sclerosis
ASI	Anthesis-Silking Interval
BCE	Before the Common (or Current) Era
BP	Before the Present
CE	Common (or Current) Era
CMS	Cytoplasmic Male Sterility
EFE	Ethylene Forming Enzymes
EM	Electron Microscopy
FAO	Food Agricultural Organization
FGISH	Fluorescent Genomic In Situ Hybridization
FISH	Fluorescence In Situ Hybridization
GISH	Genomic In Situ Hybridization
GM	Genetically Modified
GMO	Genetically Modified Organism
IRRI	International Rice Research Institute
NIRS	Near Infrared Reflectance Spectroscopy
NRCS	National Resources Conservation Service
PCR	Polymerase Chain Reaction
PPV	Plum Pox Virus
PRSV	Papaya Ringspot Vims
QTLs	Quantitative Trait Loci
RIN	Ripening Inhibitor
TMV	Tobacco Mosaic Virus
USDA	United States Department of Agriculture

PREFACE

Agriculture holds a great significance in the context of the food security and food demand of the world. The horticulture comprises a major portion of the agricultural activities and concerns the growth of crops for the further use and distribution.

The breeding of the crops is an important process to look forward to in horticulture. A lot depends on the conditions in which the plants breed properly, considering the ones that will suit them perfectly. This book attempts to bring the focus of the readers to the various processes and the other aspects related to the crop breeding.

The first chapter of the book introduces the people to the concept of crop breeding and defines plant breeding for the readers. It focuses on the genetic manipulation of the crops and informs about the way in which the plant breeding has been done over the ages. It also discusses about the genetic conservation of the crops.

The second chapter sheds light on the genetic prospective of crop breeding by discussing about the principles and the goals involved in crop breeding. It tells about the genetically modified plants and also throws some light on the use of technologies to introduce new varieties in the genetic forms of the crops.

The third chapter talks about the objectives of crop breeding and the algorithm that needs to be followed to breed the crops for good quality and yield.

The fourth chapter dwells upon the various methods that are employed for crop breeding across the globe and how they impact the process.

The fifth chapter takes the readers through the different kinds of mating systems in the crops and explains the manner in which these techniques have evolved in the plants and crops.

The sixth chapter shifts the focus of the readers to the principles and the practices that are undertaken in the breeding of the perennial crops in horticulture.

The seventh chapter takes the readers through the principles and the practices involved in breeding of the annual crops and the expected outcomes of the various processes.

And in the last chapter, the readers go through the various kinds of transgenic horticulture crops and the various aspects related to their breeding and supply.

With this book, I intend to inform the readers about the various kinds of breeding practices that are being followed around the world for crops and plants. There are many agriculture enthusiasts who are realizing the value of agriculture and its importance in the upcoming times, where the need for crops and food is bound to go up.

I hope that this book is helpful for them to undertake the studies and have information regarding the breeding of crops, and that the book can act as a base for them to take off in this field.

1

Introduction to Crop Breeding

CONTENTS

This chapter is about how the plant breeding method came into the world. This answer is given by explaining different concepts that have been brought by different scientists in this world. There are many discoveries that took place before the Mendel's experiment; those theories are explained in this chapter.

This chapter also includes concepts of genetic manipulation, and the theory of plant breeding from the ages. Its Origin, past and present. This chapter also includes a gist of contribution list. As this chapter is basically about how plant breeding method came into existence and what are development or discoveries that have been made before introducing this method.

1.1. INTRODUCTION

Plant breeding method is considered as one of the continuous and longest activity that is being conducted by the humans. The evolution of human civilization is considered that it parallels the success of the plant breeding.

Even though, it is being considered that the early plant breeding methods are simpler when compared with the present day methods. The transition that takes place from the nomadic life style of the hunter often gathers towards the sedentary life style that are considered to require the adequate supplies of the food, fuel, feed, and the fiber in order to sustain the sedentary type cultures.

It is believed that this sedentary type culture often depends on the individual to identify and to even improve the plants that are meeting their needs. It is also believed that the transition of the lifestyle was due to the availability of the plant resource towards the sustain human and livestock needs.

It is often being estimated that the dawn of the human civilization is approximately to 10,000 years ago. Therefore, it can be reasonable to the assumption that the plant breeding activities have taken place approximately 10,000 years ago, which was by selecting the plant types that was considered to be more productive and also useful to provide for the needs of humans and animals.

The initial plant efforts are considered to be necessary for the survival of the human civilization because they have developed the improved cultivars that were more productive and, in some cases, it also depends on the humans for their survival, for example, maize.

In the early plant breeder case, developing the productive cultivated plant species which was from the wild species was considered to be effective and was found to be lower in the productivity. But they are found to possess many important traits that are required for their survival in the wilderness.

Even, today, the plant breeders are occasionally seeking for the important traits like pest resistance and drought tolerance, from the progenitors of the cultivated crop species. Adaptation is one of the main traits that is being used in the selection process and found to be based on the model or ideal of the breeders.

Even though, the plant breeders have contributed a significant role in the development of different human civilization. Although the general public does not recognize this as a major contribution.

Figure 1.1: Image representing plant breeding method.

Source: Image by Flickr

Development that takes place in the transportation, medicine, space travel, electronic and engineering are considered to receive the greater attention than the plant breeding. If in case, the greater energy is needed to be allocated in order to develop an adequate food supplies that often progress in the other human activities that would have to be more limited.

For example, there are less than 2% of the population of United States that is being involved in the agricultural crop production, which often provides an adequate quantities of different crops species that can be used to meet the food, fuel, fiber, and feed needs which is a reminder of the population.

Hence, the energy and resource can be directed to other activities like space and medical research, which can also give benefit to the society. One of the main reasons behind why plant breeding has received limited credits and recognition is because there is no general understanding about what exactly the plant breeding encompasses.

Within the plant breeding discipline, it is considered that there are some differences on what define plant breeding. There have been several authors who have offered that they considered is the objectives of plant breeding.

There are some examples like: plant breeding is the genetic adjustment of plants to the service of man (Frankel, 1958); breeding is the evolution by the will of man (Vavilov, 1935); plant breeding is a unique science in at least two ways.

Firstly, there was a use of knowledge and the techniques that are from the basic science area, and secondly, its contribution towards the agricultural progress that are measured not only by the information but also by the material products like crop varieties, clones, hybrids, etc., (Frey, 1966).

The breeding that takes place in the plants and animals are the form of evolution that basically depends on the large part which is upon the same rules that often regulate the evolution of the natural selection (Briggs and Knowles, 1967).

The breeding of plants and animals are considered as an exercise that is used for exploiting the genetic system (Williams, 1964). Plant breeding can be defined as the application of techniques that is used for exploiting the genetic potential of the plants (Stoskopf et al.,1993).

Plant breeding is also an art and science that is used for improving the humankind (Poehlman and Sleper, 1995) and it is a science, business and art that is used for improving the plants for the human benefits (Bernardo, 2002), it is also used for improving the genetic pattern of the plant in relation to their economic uses.

1.2. DOMESTICATION OF THE PLANT

Domestication of a plant was considered as a man crowning achievement. Domestication often allows use to develop into the complex global society that is today. People are found to be more dependent on those same crop species that were domesticated by the early man 10,000 years ago.

It was believed that the early man joyfully embraced the new technology used in agriculture which was debatable, but once it is being caught on it which spread quickly across the continents. During speciation in the crop plants, there were many morphological changes that have been evolved in the response towards the continuous selection pressure.

Such type of characteristic was believed to be largely governed by the genetic and epigenetic changes or they are being modulated according to the ecological adaptation. The transition of a wild progenitor species into the modern elite cultivars which is through the "domestication" entailed evolutionary responses in which there is plant population adapted to the human selection.

It is believed that most of the plant species with this selection exhibits, the marked changes in the variety of phenotypes, and are most noticeable change in the trait is consciously under selection. For example, yield, fruit size, etc.

With the domestication there are also some negative aspects like reduced genetic diversity. The genetic bottle neck is being affecting which is seen in the modern crops which is a product of man's selection for the desirable agronomic traits.

It is found that the modern crops are often susceptible towards the disease, abiotic stress, and the insects. In order to find the resistance genes, it is necessary, to go back to their wild ancestors and to their close relatives.

This can behave as problematic due to the set back in the yields which is attained when the crossing to the wild relatives, but it is necessary for the advancement of the crop. The understanding of the crop domestication will help the plant breeder in the pursuit towards the next best plant.

Domestication of the plant was considered the single most important technology advance in the history. Which allowed us to develop into the highly complex civilization.

1.3. PLANT BREEDING BEFORE MENDEL

The mechanism of plant breeding occurred only informally for the thousands of years. At that time, the farmers might have chosen those method to save the seed from the healthiest or highest yielding plants that is from one generation to the next generation. But it is believed that they lack the scientific knowledge of the inheritance in order to permit the deliberate breeding for the traits or to understand the main cause and effect of the widely used method of mass selection.

Jensen, in the year 1994 of described these practices as the crop husbandry, which is distinct from the scientific attempt to improve the plants. Therefore, it is important to remember it without any written records of their methods. The point which is known is that the development of the land laces and prehistoric agricultural practices often comes from the gleanings of the archeological evidences.

Even though it is not known exactly when the agriculture was invented, Mangelsdorf (1952) often considered that based on the archeological dating of the oldest cultivated plant remains, it has been recent. For example, there are most of the investigation that has been done into the domestication of maize which is a vital crop.

The oldest example of such investigation was found to be in the Tehuancan Valley of the central Mexico, dating back 7,000 years back (Jensen, 1994). Even though, it was found that the maize is related to the teosinte, where there were still widely divergent opinions for many years which was regarding from what exactly the maize was domesticated .

Near about six decades ago, Randolph, 1952 was found to be tentatively willing to commit to Mexico or central America as the domestication origin of the maize. Randolph felt that there is a possibility of Asiatic domestication which is via emigration through the Bering Straits, that should not be discarded.

There are many other domesticated plants that have experienced similar drastic changes during the comparatively brief time period which as a result, only a few hundred or thousands of years of efforts by the humanity (Mangelsdorf, 1952).

Later, Mangelsdorf reminds that the contrary to the Darwin's beliefs, the mutation is the power evolution in the domestication – where domestication itself does not produce any new inheritable variations. In early example of the plant breeding, it has been opposed to the crop husbandry which basically began in the early part of the 17th century.

In that century, the pilgrims from the Europe to the new world often discovered that there are traditional crop varieties that were eminently considered to be unsuited to their new homes. Kloppenburg, in 2004 often described about how the initial settlers have been obtained and how they grew Native American varieties for the sustenance while slowly adjusting their own varieties and species to the stock.

During early days of American agriculture, the farmers use to select the best crop plant from each year's crop in order to provide the good quality seed for the next generation and sometimes they also acquire the seed from other farmer and migrants by engaging in both the introduction of the new germplasm and in simple mass selection (Kloppenburg, 2004).

Although, it is considered that they were not selecting for the specific traits or they are using the scientific methods, their obligated labors for the survival that resulted in the adaptation of the varieties which is possible via basic plant breeding method.

There is even more advance development in the plant breeding that would have been made impossible with the initial scientific forays of the 17th century. It is believed that two of the plant anatomists, namely MarcelloMalpighi and Nehemiah Grew, each of them had made the substantial contributions towards the early understanding of the plant anatomy and reproduction (Ingensiep, 2004).

By using these novel innovations of the microscopy, Malpighi often described the plant structure and the generation of the seed in the 1675 publication, then in the 1679, elaborate on the germination and growth of the young plants.

Event thought the flowers sexuality was still an unknown concept at this time, and Malpighi used the considerable background in the animal anatomy and medicine in order to give the names to some of the reproductive parts of the plants (Ingensiep, 2004).

Although, there are many gardeners in the late 17th century who believed that the seed production often involves pollen, but for most of them felt that it was just a philosophical conjecture. In the year 1694, Rudolph Camerarius was able to prove this, and even demonstrated that the necessity of the pollen through the series of systematic experiments which emasculated the flower from the variety of the species and some of them are hand pollinated (Thompson, 2010).

When there was production of only pollinated flower seeds, then Rudolph was able to conclude that the anthers which are often being producing the pollen can be male and the ovaries, stigma and style might have received the pollen and must be the part of female.

Even though it was noticed that the Camerarius' discoveries were largely being ignored for the several decades. Which might be possible due to his colleagues who did not believed to the belief that the plants are asexual, as in the Thompson research. The Thompson research served as a stepping stone for the Joseph Koelreuter's investigation in pollination seventy years later.

By following the anatomical investigation of the 17th century, in the 18th century there was an increase in the focus on a relatively with the new innovation in the agriculture which is basically the hybridization of one kind of plant with another.

In the year 1717, the earlier successes that takes place with the technique was performed by Thomas Fairchild who basically worked on crossing between the carnation and a sweet William that often resulted in a flower that is known appropriately as a Fairchild's Mule (Thompson, 2010).

There were so many researches like continuing Camerarius performed research in pollination, Joseph Koelreuter performed on a systematic series of the scientific experiments that was on the hybridization, determination on whether the characteristics in the plants are always being inherited from the mother or from the father only.

Working even with the wide varieties of the flower species and tobacco, Koelreuter demonstrated that the hybrid offspring often displays the characteristic that is from both the parents, generally in an intermediate form.

Performing crosses additionally from the hybrid tobacco and dianthus plant back to their own parents, Koelreuter was found to be able to show that over many generations the progeny is being resumed from the phenotype of the recurrent parents (Bailey, 1906). This process is basically a foundation of the backcrossing method, which is extensively used nowadays.

Same as Koelreuter, Thomas Andrew Knight often performed on the amount of the plant breeding during the 18th century who was as a gentleman farmer without the benefits of a formal scientific education.

Thompson, 2010 often described that knight's curiosity was regarding the wide variety of the topic that was related to the plant growth and production which led them to perform the process of hybridization on various fruits and vegetables which generate the improved cultivars.

Even after creating the two hybrid varieties of strawberry, in the year 1787, Thompson often began experimenting with the crossing inbred lines of the sweet pea plant. It was basically foreshadow of the Mendel revolutionary studies with the same species by using the better part of the century (Thompson, 2010).

Even Thompson was unable to explain the dominant and recessive alleles, Knight (1809) could practically shows the breeding results and demonstrate building blocks of what exactly the science of the genetics, even without having a scientific background.

It was just a year before, when Knight began to work with the peas, Franz Achard also initiated some breeding programs for the sugar beet which operated from 1786 until 1830. By simply performing the mass selection on the fodder beets for increasing the sucrose content. Achard and Von Koppy family to create created the new cultivars with high levels of sucrose.

This helped in improving the specific quality of a fodder crops which served to elevate it into a food crop resulted in the creation of the major industry and extracting of the sucrose from sugar beets (Dudley, 1994).

In the 19th century there was continuation of the trend of organizing the efforts in the plant breeding which was being performed by various groups and organizations (Jensen, 1994). It was having some exceptions as various curious and scientifically minded individuals were found to have a historical background in this case.

In the year 1906, the edition of the classic text on the plant breeding, Liberty Hyde Bailey described some of the recent plant breeding practices including the effort of the Luther Burbank. Bailey was considered as a practical plant breeder for the commercial purposes.

During that year Burbank was responsible as a plant breeder for creating near about 800 varieties of the plants which also included a spineless cactus and intended for the livestock fodder, the Santa Rosa plum, the Freestone peach, and the Burbank and Russet Burbank potatoes (Thompson, 2010).

Bailey in 1906, described Burbank as a genius and preeminent plant breeder of their time and praised him for both the dedication to the plant breeding and the kind nature. In addition to the Burbank and to the commercial plant breeders there are much work done at the experiment stations and agricultural colleges.

It was not only in research and variety improvement, but also with the high quality seed and of its ideal conditions (Bailey, 1906).It was noticed that the sheer scope of these works and the manpower behind them is allowing them to yield the result in the untold number of projects.

Similarly, Bailey, 1906 credited the USDA "the largest organized governmental plant-breeding enterprise in the world" which have been created in the year 1862 and continued the American government tradition of collecting and disseminating the germplasm to growers (Kloppenburg, 2004).

French Vilmorin Company was considered as one of the foremost seed companies in that period which was founded in the 18th century by the de Vilmorin family. In the beginning of the 1840 and for the four decades, it was found that the company was focusing on the original method based on selection of plant lineages, rather than the individuals which can be later known as a "genealogical selection" (Gayon and Zallen, 1998).

The Vilmorin Company has also enjoyed a lot of economic success with the wheat and other crops through the methodical combination of the hybridization and selection which often begun in eth 1870s (Gayon and Zallen, 1998).

It is considered that the member of the de Vilmorin family are having a significant influence in the realm of the plant breeding over the span of ore than the century, not only by the variety of creation and improvement, but there are more abstract philosophical contribution towards the filed (Bailey, 1906).

1.4. DEFINING PLANT BREEDING

It is considered that plant breeding is a deliberate effort made by the humans in order to nudge the nature, which is basically with respect to the heredity of the plants and to the advantages. These changes which are made in the plant are considered to be permanent and heritable.

Therefore, the professionals who are basically conducting these tasks are called as plant breeders. These efforts are considered to adjust the status of the quo which is often investigated by the desire humans in order to improve some certain aspects of the plants used to perform the new roles or even enhance the existing ones.

Although, the phrase "to breed plants" is often considered to connotes the involvement of the sexual process which basically effecting the desired changes. It is observed that the modern plant breeding is also including the manipulation of the asexual reproducing the plants, especially those plants that do not reproduce through the sexual process.

Breeding is therefore about manipulating plant attributes, composition, and the structure in order to make them even more useful to the humans. It is important to mention that at the onset it is not every plant characteristics or traits that is basically amenable towards the manipulation by the breeders.

However, as the technology advances, the plant breeders are also increasingly able to accomplish the astonishing plant manipulations basically the case involving the development and applications of the biotechnology towards the plant genetic manipulation.

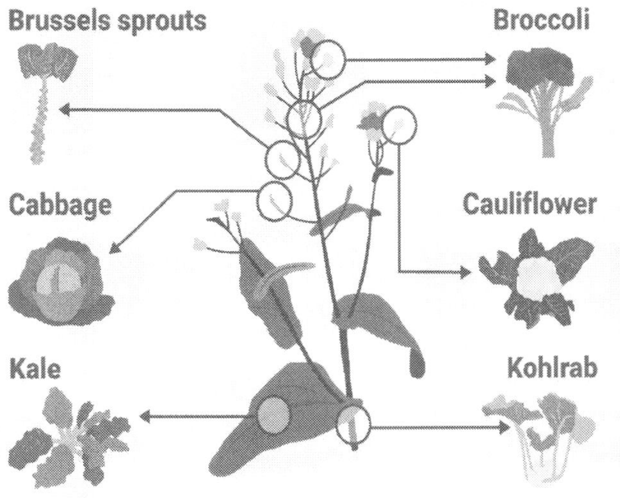

Figure 1.2: Image of wild mustard plant an example of plant breeding.

Source: Image by Business Insider

Transgenesis is considered as one of the most controversial modern technologies. It is basically the technology through which the genes are transferred and is made across the natural biological barriers. Plant breeders are found to be specialize in breeding the different groups of the plants.

There are some who focusing on the ornamentals, forage crops (e.g., alfalfa, grasses), field crops (e.g., soybean, cotton), turf crops, fruit tress (e.g., citrus, apple), horticulture crops (e.g., vegetables). It is more important that the breeders are focusing on the specific species in these groups.

This is the way through which they are able to develop the expertise that often enables them to be the most effective for improvising the species of their own choices. The principles and the concepts are being discussed in this chapter and generally it is applicable towards breeding all the species. Therefore, most of the examples are being supplied are from the breeding filed crops.

Plant breeding is consisting of four main phases, where all of them are running concurrently in an established program. These programs are:

1.4.1. Phase 1

It is observed that the breeders are identifying the needs of the farmers and the deficiencies that are in the current varieties. However, it is important to improve or to find the new disease resistance, or increase the size of the seed or simply increase the yield in order to make the crop profitable.

Then the breeders are needed to collect together the separate genotype that is basically having the attributes that is required. This may also require the screening that is available of the germplasm collection or obtaining it form the other breeder's genotype that is described in the scientific literature.

If the availability of the attribute that is often required is not present in the species gene pool than it is considered that the breeders may start using the gene technology in order to obtain the gene from somewhere else or as a last resort, considering of synthesizing a completely new gene.

Latter there are two options that will be long term strategic decision that is depending on the values of the overcoming the deficiency and also the facilities and collaborators availability.

1.4.2. Phase 2

According to the second phase of the breeding program the main aim is to artificially hybridize or cross the identical parents and to bring the genes for the desirable attributes, together in the same hybrid individual.

It is believed that the practice procedure will depend on the species involved and any of the pre-existing knowledge of the genetic control of the attribute.

1.4.3. Phase 3

Earlier, in the segregating generations the breeders are selecting the progeny of the crosses which helped them removing undesirable or inferior genotype, which is progressively moving toward the small number of the elite lines.

The third phase is considered as a largest part of the breeding program which involves the identification of the product of a genetic segregation and recombination and even finding the "best of the brunch" as a reliably and quickly as possible. Meanwhile, minimizing the risk of failing to retain the superior lines.

1.4.4. Phase 4

The final phase of the breeding often consists of establishment which is worth of a new genotype over the existing varieties, with that bulking of the sufficient quality of the seed for the distribution among the farmers and finally releasing the new variety.

It is considered that the last phase also consumes a significant amount of the breeding resource, even though there is small number of advances in the program each year. It is important to evaluate in an extensive field trial program at many locations, and are also having large quantities of produced.

Breeder are constantly facing the dilemma of having the varieties released as quickly as possible. Even they are still having some reliable data present on the variety regional performance and it is performing in a range of years.

Breeders are actually aware about the farmers need that are convincing the data to change the variety and also feel a strong responsibility to provide the information as accurate and is as complete as possible.

1.5. CONCEPT OF GENETIC MANIPULATION OF PLANT ATTRIBUTES

It is considered that the work of the Gregor Mendel further advances that took place in the science are followed by the Mendel's discoveries establish that plant characteristics are being controlled by the hereditary factors or genes consisting of DNA which is a hereditary material.

These genes are being expressed in an environment to produce a trait. It basically follows to change the trait or either its expression, it is necessary that one may change the nature or genotype or can even modify the nature.

Changes in the environment is essentially entails the modifying of the growing or production conditions. An organic approach can achieve this, for example, the applications for the production inputs like irrigation and fertilization.

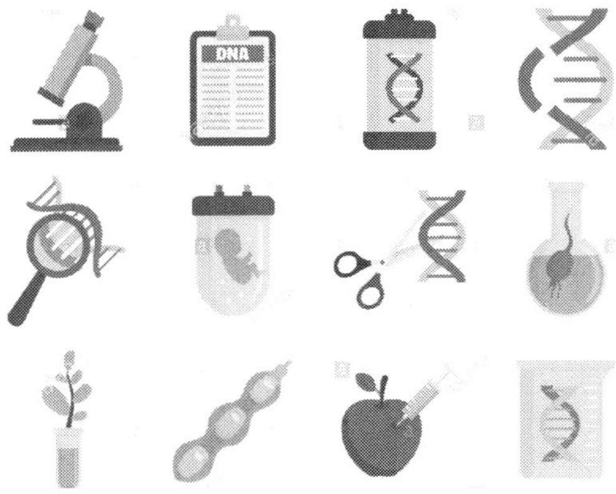

Figure 1.3: Representation of genetic modification.

Source: Image by Alamy

Therefore, this approach is considered as effective which is used for enhancing some certain traits and the facts that basically remains supplemental environmental factors are being removed and the expression of the plant traits are being reverted to the status quo.

Whereas, on the other hand the plant breeders are seeking for modifying the plants to the expression of certain attributes by modifying the genotype in a desired way by simply targeting the specific genes. Such approaches are believed to produce an alteration which is permanent, which means transferable from one generation to another generation.

1.6. PLANT BREEDING THROUGH THE AGES

1.6.1. Origins of Agriculture and Plant Breeding

In the primitive form, the plant breeding started after the invention of the agriculture. Basically, when the people from the primitive culture changes their lifestyle from the hunter gatherers to the sedentary producers of the selected plants and animals.

It is observed that the agricultural origin basically range from the mythological to ecological. These changes in the lifestyle do not occur overnight, but it was a gradual process which takes place through which the plants are transformed from being independent, fully dependent on humans, domesticated varieties, and wild progenitor.

Agriculture is generally found to be as a discovery and invention. During this period, humans are also found to discover the time-honored and as the most basic plant breeding technique which includes selection, pick desirable variations and the art that is used for discriminating among the biological variation in the population to identify and pick the desirable variants.

Selection often implies the existence of the variability. In the beginning of the plant breeding it is the variabilities which are exploited, the naturally accruing variants and the wild varieties of the crop species.

Further the selection was on the basis of exclusively on the judgement, intuition, and the skill of the operator. In the poor economies, this form of selection is practiced today by the farmers, where basically they save the seed from the best-looking plants or from the most desirable fruit for the purpose of planting the next generation or season.

These days, it is found that the scientific techniques are being used to the quantities known before for making the selection process more precise and efficient.

Although, the activities are described in here are found to be similar to some of those which are practice by the modern plant breeders. It is not suggested that the primitive crop producers were necessarily conscious of the fact that they are nudging the nature towards the advantage as a modern breeder doing.

1.6.2. Plant Breeding Past (Pre-Mendelian)

It is believed that the early plant breeders do not purposely created the new variants. Whereas the modern plant breeders are found to be able to create the new variants that is previously did not occur in the natural population.

It is difficult to identify the true beginning of the modern plant breeding. However, there are certain early observations that are made by some certain individuals which helped to lay the foundation especially for the discovery of the modern principles of the plant breeding.

It has been reported that the archeological records are indicating that the Babylonians and Assyrians are artificially pollinated by the date palm, which is at least 700BC. R.J. Camerarius also known as Rudolph Camerer, who was from Germany was the first who was credited with the reporting of the sexual reproduction in 1694.

R.J. Camerarius'experiments shows that the pollen from the male flowers is indispensable towards the fertilization and seed development on the female plant. Camerarius conducted his work on the monoecious plants like maize and spinach.

However, the first person who does systematic investigation into the plant hybridization of a number of species was Joseph Koelreuter, between the years 1760 and 1766. Similarly, Thomas Fairchild who was an Englishman, in the year 1717, conducted an interspecific crop which means the cross that takes place between the two species, between the sweet William (*Dianthus berbatus*) and *D. caryophyllus*.

This cross was basically done in order to what became known as Fairchild's Sweet William. There is another account that often describes the observation in 1716 by an American, Cotton Mather, in order to affect that ears from the yellow corn grown next to the blue or red corn which had a blue and red kernel in them.

This often suggests the occurrence of the natural cross pollination. Maize is considered as one of the crops that has received an extensive breeding and genetic attentions in the scientific community. In the early 1846, Robert, who was from Illinois was being credited with the developing what basically known as the "Reid Yellow Dent".

This landmark is considered to work by the Swedish botanist, Carolus Linnaeus from 1707 to 1778, which culminated in the binomial systems of the classification of the plants which is invaluable to the modern breeding of plant.

In the year 1727, Louis Leveque de Vilmorin who was of the Vilmorin family of seed growers which founded the Vilmorin Breeding Institute in the France as their first institution which is dedicated to the plant breeding and to the production of the new cultivars.

There are also another commonly used breeding techniques like – progeny test which means growing of the progeny of a cross for the purpose of evaluating the genotype of the parent. This progeny test was first used to evaluate the breeding values of a single plant.

1.6.3. Plant Breeding Present (Post-Mendelian)

Method of modern plant breeding basically depends on the principle of geneticist which is the science of heredity to which Gregor Mendel has made some of the foundational contributions. Mendel's is found to originally work on the garden pea which was published in the year 1865.

Mendel work described about how the factors are for specific traits that are being transmitted from parents to the offspring and through the subsequent generations. Mendel's work was rediscovered in the year 1900 with the confirmation given by E. von Tschermak, C. Correns and H. de Vries.

These events laid the foundation of the modern genetics. Mendel's studies often gave birth to the concept of genes and the discipline of the genetics, factors that are found to encode the traits and transmitted through the sexual process towards the offspring.

Mendel's work often resulted in the formation of the rules that are important for understanding the heredity that are caked as Mendel's laws. Danish botanist, the Wilhelm Johannsen was the first to tell the earliest application of genetic plant breeding.

In the year 1903, Johannsen developed the theory of pure line while working with the garden beans. This work was confirmed as an earlier observation by others and the technique of selection can be even used to produce the true and uniform breeding cultivars.

These are being developed by simply selecting it from the progeny of a single self-pollinated crops which through the repeated selfing to obtain the highly homozygous lines which are later crossed. Earlier, H. Nilson had demonstrated about the unit of selection of the plant.

The product achieved by the hybrid plants is being performed by parent with respect to the trait of interest. Hybrid vigor or heterosis is basically the foundation of the modern hybrid crop production program.

In the year 1919, D.F. Jones took the idea of a single cross further by proposing the concept of double cross. This concept basically involves the cross that takes place between the two single crosses. This technique helped in making the commercial production of the hybrid corn seed economical.

The application of genetics that is used in crop improvement has been observed to yield the spectacular success over the year, and one of the most notable development of the dwarf and producing the environmentally responsive cultivars of rice and wheat especially for the subtropical region of the world.

It is considered these new plant materials are transforming the food production in these regions which is in a dramatic fashion and process, it became the dubbed Green Revolution. Mutagenesis is basically the induction of the mutations by using the mutagenic agents like chemical or radiation.

Mutagenesis has become a technique used for the plant breeding in 1920s, at the time when the researchers discovered that exposing of the plants to the x-ray increased the variation in the plants. Mutation breeding was accelerated after the World War II when the scientist often includes the nuclear particles like protons, alpha and gamma, as a mutagen in an organism.

Although there are very unpredictable in the outcome, there is a successful use of mutagenesis that has been developing numerous mutant varieties. In the year 1944, DNA has been discovered to be the genetic material.

Later, the scientists began to understand the molecular basis of the heredity. Few new molecular tools have been developed in order to facilitate the plant breeding. Recently, scientist found to be able to circumvent the sexual process used for transferring the genes from one parent to the other.

The genes can be transferred from any organism to another. Genetic engineering is considered as a new tool which has its proponents and distractions. Currently, the success includes the development of the insect resistance which is in the crops like maize by which the incorporating of a gene from the bacterium *Bacillus thuringiensis*.

Cultivars are considered to contain an alien gene specifically for the insect resistance from this specific organism that is called as BT cultivars, which is a diminutive of the scientific name of the bacterium. Therefore, the product of the application of the specific alien gene transfer technology is generally known as a genetically modified (GM) or transgenic product.

Plant biotechnology is basically the umbrella name used for the host of the modern plant manipulation techniques that might has produced among other things, molecular marker is consist to facilitate the selection process in the plant breeding.

1.7. GENE POOLS

Presently it is believed that the crop breeding is drawing usefully on the genetic variability that is not only in the crop species but also from the related genera and species. One of the examples that can be used for the distantly related genetic variability in the crop breeding is basically the incorporation in the wheat gene for resistance to the stem rust, (*Puccinia graminis tritici*) are from the *Agropyron elongatum* (Knott, 1963).

These species are considered to be distantly relative of the wheat, but it is found to be able to hybridize with the wheat and possessing the genes that can be useful to contribute towards the genetic improvement of the crop species.

Figure 1.4: Representation of gene importance.

Source: Image by Pixabay

`It is believed that the concept of the gene pool of the species is believed to embrace the genetic variabilities within the cultivated forms. It is considered as a weed and wild relatives and is found to be more distantly related with the species which can be hybridized with the commercial species and even can contribute to the genes for its improvement.

Harlan et al. (1973), recognized that the gene pool of a species is considered to have a primary and secondary levels that are commonly related to the degree of the accessibility of genetic variability for the crop breeding within the closely and more distantly related species.

It basically involves the consideration like hybrid variability, chromosome pairing and the cross compatibility in the hybrids with the crop species. It is found that the modern techniques in the genetic engineering or molecular biology is having a great extent of the concept of a gene pool in order to include any kind of gene in the existence or any of the gene that can be able to construct and design in the laboratory.

Novel gene or Alien genes are used in the plant breeding which bring them potentially a huge benefit. But it is found that it also ranges a fundamental difficulty like reliable inheritance, those of integration and expression.

Therefore, these technical difficulties are believed to overcome in the time and transgenic varieties that will help in increasingly be considered as a normal part of the potential gene pool. There are many people who are ethically and morally opposed towards the transgenic organism and therefore the probability of their wide acceptance by the consumers is considered to remain unclear.

1.8. GENETIC CONSERVATION

The use of life blood of crop breeding for the purpose of improving the disease resistance, quality, further improvement in the yield and other characteristics is considered as the genetic variability that is available within the gene pool of the species.

Therefore, in order to retain the maximum flexibility in the breeding and the ability that is used to respond to new threats and challenges in to the crop is considered as necessary in order to have the availability as a wide range of the genetic variability as possible.

As there are many of the long term future crops that breeding needs and are unknown as it is considered as a necessary genetic variability which is regardless of its immediate value, and is being conserved for the future crop improvement.

It is considered that most of the original genetic variability of the crop is located in area of primitive agriculture. It is believed that the implementation of these agricultural improvement and development programs that takes place in these areas created much of the variability of the primitive varieties. It is found that the associated weed species to be lost which is due to the widespread cultivation that takes place in new and uniform varieties from the overseas.

It is found that the breeders and geneticists are much aware about the needs to conserve the genetic variability for the future crop breeding (Frankel and Hawkes, 1975). The Food and Agricultural Organization (FAO) in United Nations is find to fund much of the maintenance and collection of the genetic variability of the crop plant and related species, along with the international breeding centers in the CHIAR network.

There are number of countries who are maintaining the gene bank or the germ-plasm collection of many of the world's crop plants. There are basically large collection of the advanced varieties and crossbreeds, wild and weed relatives, primitive landraces, and other species of the crop plant.

United States Department of Agriculture maintains a large gene bank specifically for the crop plants which is present at the Beltsville Maryland, and Russia which has a similar collection at the N.I. Vavilov All - Union Institute of Plant Breeding at the St Petersburg.

In Australia, a national winter cereals collection has been established at the Tamworth, New South Wales, basically for the use by the Australian breeders. The Germplasm collection related to crop plant and their wild relatives is considered as an important for the Australian agriculture held at CSIRO in Canberra, Biloela in Queensland, Perth in Western Australia, and Horsham in Victoria.

In the gene bank the storage is mainly used as seed which is kept under the condition of low temperature and humidity in order to ensure the viability for 10 to 20 years. Interest have been shown in the possibility for storing the pollen of the crop plants in the gene banks. But it is found much shorter viability than the stored seed.

It is considered that the pollen collection is much more difficult to regenerate than seed collection, a factor which is a serious limitation to provide facility of maintaining the viability of the lines in a gene bank.

Before the viability of the declines towards the low level, the seed collections which are sown and are harvested to produce the fresh seed and in this way the gene bank gets opportunity to be perpetuated.

It is believed that some of these collections are extended to accumulate the vegetative propagules like single cell cultures or tissues, especially those species whose seed or pollen is not easy to be stored which is due to rapid loss of the viability.

During the Phase 1 of the breeding program, the breeders are electronically searching for the germplasm collections of the species that are concerned for the accessions that can contain the genes that are required to start the crossing program, the seed sample are then procured.

1.9. AN INTEGRATED AGRICULTURAL MODEL

It is noticed that the above-mentioned scenario can be tackled only by introducing an integrated agricultural model. This means that combining of the best of the conventional agriculture with the best ecological insights from the organic agriculture, which is together with the adoption of the latest technologies and even findings.

Figure 1.5: Genetically modified crop.

Source: Image by Pixabay

This applies both to the methods of plant genetics and cultivation. In the dossier there are different methods by which many of the humans are adapting the genetic traits of the crop from the ancient times. But there is a discussion of new techniques that have emerged recently, and will be used in the future.

There is a focus on the range of the New Breeding Techniques that is getting an extra political and public attention in context of GMO debate. This is done by focusing on the plant breeding to detract from the value of cultivation insights and cultivation techniques.

These are considered as important aspect in increasing and sustaining the food production and it goes hand in hand with the plant breeding.

1.10. SELECTED MILESTONES IN PLANT BREEDING

- Around 9000 BC the first evidence of the plant domestication has been found in the hills above the Tigris river.
- In 3000 BC, domestication of all the important food crops takes place in the Old World.
- In 1000 BC the domestication of all the important food crops takes place in the New World.
- In 700 BC the Assyrians and Babylonians worked on hand pollinated date palms.

- In 1694 the Camerarius demonstrated that sex in the plants and even suggested the crossing as a method to obtain the new plant types.
- In 1716 the Mather of USA observed the natural crossing in maize.
- In 1719 the Fairchild created the first artificial hybrid of the dianthus sp. (Carnation × Sweet williams).
- In 1727, the Vilmorin introduced the concept of the progeny testing i.e., known as Vilmorin isolation principle.
- In 1753, the Linnaeus published the species of plantarium and the binomial nomenclature was born.
- From 1761 to 1766 Koelreuter of Germany often demonstrated that the hybrid offspring receives the traits form both the parents and were intermediate in most of the traits. They produce the first scientific hybrid by using the tobacco.
- In 1847, development of "Reid Yellow Dent" was developed.
- In 1866, Mendel published his discoveries in the Experiments in plant hybridization and cumulating inheritance and discovery of the unit factors like genes.
- In 1899, Hopkins described the ear-to-row selection method of breeding in maize.
- From 1908 to 1909, Hardy of England and Weinberg of Germany developed the law of the equilibrium of the populations.
- In 1908, Nilsson Ehle proposed the concept of multiple factor hypotheses.
- In 1909, Shull conducted extensive research to develop the inbred to produce the hybrid of maize.
- In 1917, Jones developed the first commercial hybrid maize.
- In 1926, Pioneer Hi-bred corn company established as the first seed company.
- In 1934, Dustin discovered colchicines.
- In 1935, Vavilov published the scientific basis of the plant breeding.
- In 1940, Harland used the bulk breeding selection method in breeding.

- In 1944, Avery, McCarty and McLeod discovered the DNA which is basically a hereditary material.
- In 1945, Hull proposed the recurrent selection method used in breeding.
- In 1950, McClintock discovered the Ac-Ds system which is of the transposable elements.
- In 1953, Watson, Crick, and Wilkins proposed a model for the DNA structure.
- In 1970, Borlaug often received the Nobel Prize for the Green Revolution.
- Berg, Boyer, and Cohen helped introducing the recombinant DNA technology.
- In 1994, the development of "Flavr Savr" tomato is developed as a first genetically modified food which is produced for the market purpose.
- In 1995, BT corn is developed.
- In 1996, Roundup Ready often introduced soybean.
- In 2006, Roundup Ready developed the wheat.

1.11. FUTURE OF CROP BREEDING

As it is expected that the world population is expected to increase continuously, with that there will eb continuous demand of food. However, with the increase in population, there will be increase in the demand for the land for the residential, reactional, and commercial purpose.

Figure 1.6: Image representing crop breeding method.

Source: Image by Wikimedia Commons

Many times, it is observed that the farm lands are converted for the use. Increasing demand of the food production can be achieved by increasing the production per unit area or even bringing the new lands into the cultivation.

There are some ways through which the society can affect and can be affected by the plant breeding in the future which are as follow:

1.11.1. New Roles of Crop Breeding

Traditionally the roles of the crop breeding like for feed, fiber, food, and the ornamentals will continue to play important roles. However, the new roles are found to be gradually emerging for the plants. The technology that is for using the plant as a bioreactor in order to produce the pharmaceuticals will get advance and this technology will remain around the world over a decade.

Strategies are considered to be perfected in order to use the plants to generate the pharmaceuticals antibodies, altering the plant phenotype by immunomodulation, and by engineering the antibody mediated pathogen resistance.

There have been several successes achieved like incorporation of the Streptococcus surface antigen in the tobacco plant and the herpes simplex virus in rice and soybean.

1.11.2. New Tools for Plant Breeding

It is believed that there will be a development of new tools for the plant breeders, especially in the areas of the application of biotechnology to the plant breeding. There will be the development of new marker technology and the older ones will get advanced.

There will be enhancement of tools that will basically assist the breeders towards more effective manipulating quantitative traits.

1.11.3. Training of Plant Breeders

It is believed that as there is an increase in the role of biotechnology in plant genetic manipulation, therefore graduates who combine their skills and knowledge in both the molecular and conventional technologies are in high demand.

It has been observed that there are some commercial plant breeding companies who are giving chances to graduates with the training in the molecular genetics and then they are providing them job with the need of plant breeding skills.

1.11.4. The Key Players in Plant Breeding Industry

In the last decade it was seen that the fierce race by the multinational pharmaceutical corporations are acquiring the seed companies. There are several key mergers. The modern technologies of plant breeding are basically concentrated in the hands of these giant companies. The trend of acquisition and mergers is likely considered to continue in the future as well.

1.11.5. Yield Gains of Crops

With the increase in the policing of the environment and with the dwindling of the arable land by the activist, there is an increase in the need in order to produce more food or other crops products that are on the same piece of the land in a more efficient and environmentally in the safer manner.

It is believed that the high-yielding cultivars will continue to develop, especially in the crops that might have received less attention from the plant breeders. Breeding that is based on adaptation of the environmental stresses like salt and drought, that will often enable more food to be produced on the marginal land and will even continue to play important part.

1.11.6. The Biotechnology Debate

It is believed that these modern technologies especially for the plant genetic manipulation benefits for most of the developing countries as they are in dire need of food, which is both in quantity and nutritional value.

Whereas, on the other the intellectual properties which are covering these technologies is basically owned by the giant multinational corporations. Efforts are being made continuously for making the negotiate fair use of these technologies.

When there is appropriate transfer of technology and support for the poor third world nation which continue enabling them to develop the capacity especially for the exploitation of these modern technologies.

1.12. CONCLUSION

Plant breeding is basically an application of a genetic principles to produce the plants that are more useful for the humans and animals. This method is accomplished by the selection of plants found to be more economically or aesthetically wanted. This is done by controlling the mating of individual those are selected and then selecting of certain individuals among the progeny.

This process is being repeated over many generations and is believed to change the hereditary value of the plant population which are far beyond the natural limits of the existing populations. There are basically multiple traits that are important for selecting the species in the plant and animal breeding.

Therefore, in order to focus on those specification to change or develop into new breeding variety it is important to first understand how all these achieved. This simply means that in order to have these changes in our day to day life and to fulfill all the need of the increasing population, it is necessary to understand how differently scientists have worked on different projects and tried to bring their result into the real world.

There is different concept bought by different scientists, that is combined together to form the basis of genetics. Like work of Gregor Mendel, who explained two principles which played an important role in the inheritance. Similarly, there are many who worked on different projects and bring a clear definition of genetic and its roles.

REFERENCES

1. 2012. *Introduction to Plant Breeding*. [ebook] Available at: <https://blogs.extension.wisc.edu/mgvfruitveg/files/2012/11/plant-breeding-1-up.pdf> [Accessed 9 June 2020].
2. 2016. *The Past, Present and Future of Plant Breeding*. [ebook] Belgium: VIB. Available at: <http://www.vib.be/en/about-vib/Documents/vib_facts_series_fromplanttocrop_ENG.pdf> [Accessed 9 June 2020].
3. Brown, N., 2015. *2. History of Plant Breeding - Plantbreeding*. [online] Plantbreeding.coe.uga.edu. Available at: <http://plantbreeding.coe.uga.edu/index.php?title=2._History_of_Plant_Breeding> [Accessed 9 June 2020].
4. Ecoursesonline.iasri.res.in. 2012. *GPB 301: Milestones in Plant Breeding*. [online] Available at: <http://ecoursesonline.iasri.res.in/mod/page/view.php?id=7648> [Accessed 9 June 2020].
5. Hallauer, A., 2011. *Evolution of Plant Breeding*. [ebook] Crop Breeding and Applied Biotechnology. Available at: <https://www.scielo.br/pdf/cbab/v11n3/a01v11n3.pdf> [Accessed 9 June 2020].
6. Jagtap, D., n.d. *PLANT BREEDING AND SEED TECHNOLOGY*. [ebook] Available at: <http://studymaterial.unipune.ac.in:8080/jspui/bitstream/123456789/4932/1/Plant%20Breeding-Introduction%2C%20Scope%20and%20Importance.pdf> [Accessed 9 June 2020].
7. Luckett, D. and Halloran, G., n.d. *PLANT BREEDING*. [ebook] Available at: <https://cdn.csu.edu.au/__data/assets/pdf_file/0007/2805559/Chapter4_LuckettHalloran.pdf> [Accessed 9 June 2020].
8. Miflin, B., 2000. *Crop Improvement in the 21St Century*. 51st ed. [ebook] Available at: <https://watermark.silverchair.com/510001.pdf?token= AQECAHi208BE49Ooan9kkhW_Ercy7Dm3ZL_9Cf3qfKAc485ysg AAAnMwggJv BgkqhkiG9w0BBwaggg Jg MIICXAIBADCCAlUGCSqGSIb3 DQEHATAeBglghkg BZQMEAS4wEQQMBb928gwJxGw8gs8N AgEQgIIC JuvZ4OrPV8S9_ZyB15uk2Pcwc5y7jOl14h MzpWprs3tuUuC8wy5zd6gKKOuaZR3cSlYSadKCak nER5eTLoc47-TyDi9CXXJmjhF fW1tpPEZka Dupg2 L1Lnaqx5j33s6x4WWKqb4kVKaIJGWmZZbXf5HN2YP4Rd9 Emigl FqzHjRyr2hQQOZsXqHbK MlghsL gX0B70X-6weXKlS8Qq24os NZWR1RbgHSSkLnIsA3_UN8o3HEQ7Ky

BjqcFmobmhekMsxTqmDuR0TQ-iK_VDDFKgBH64WrBzQRkKTiXfGyS9AB7O3gWlWiibRqZCuYRQvK7Od6FrDaikMBGZ3-tZly4rVbpmIS8LZC5CKOpE8XAvKfXTSsOF-GB9Jt8A_GROeJ7b1aR4VgOOjAszAFbSxNwF3GA2NPECf5D_y0-RjshahJLDgmIlagLvX467q_eD4nRJ-IzBODnNcEGguYWgr5-2vpHkZagfNZj7Qo3BnY-XxzM9TjElWHyALFN7DGP4BM2-JEuGb51t7M60wIjCmLGDgGLiEdw5ibIFV0q2SX91B2pljhhw1QOvT6nECk50sqzR8yoO7Yiq6sqZjKW_4zjMm8kr0KiKlnBcvHIEq0zZbcP7pw2ajxkj8MUjZGPRm-wQW20h_ex5n2WZwY1DQqkadprJc9QrWacz6468mzi26-TaOuprikNZr_t2snXAC2w2RBtArg3xx4Fh5Rr_OmzXCWvqCvNyhbg> [Accessed 4 June 2020].

9. n.d. *History and Role of Plant Breeding in Society*. [ebook] Available at: <http://agro.unc.edu.ar/~mejogeve/History%20and%20role%20of%20plant%20breeding%20in%20society.pdf> [Accessed 9 June 2020].
10. n.d. *Introduction to Plant Breeding*. [ebook] Available at: <https://jvanderw.une.edu.au/monday.pdf> [Accessed 9 June 2020].
11. Bailey, 1906
12. Bernardo, 2002
13. Briggs and Knowles, 1967
14. Dudley, 1994
15. Frankel and Hawkes, 1975
16. Frankel, 1958
17. Frey, 1966
18. Gayon and Zallen, 1998
19. Harlan et al., 1973
20. Ingensiep, 2004
21. Jensen, 1994
22. Kloppenburg, 2004
23. Knight, 1809
24. Knott, 1963
25. Mangelsdorf, 1952
26. Poehlman and Sleper, 1995
27. Randolph, 1952
28. Stoskopf et al.,1993

29. Thomas Fairchild, 1717
30. Thompson, 2010
31. Vavilov, 1935
32. Williams, 1964

2

Genetic Prospective of Crop Breeding

CONTENTS

This chapter is about the crop breeding methods that are involved for developing changes in the offspring. This chapter gives a brief discussion about the Mendel's law and many other experiments that have been performed for studying the genetics of the crop and to bring further changes in the crops.

Then further, new genetic variation that were incorporated in the field of science and technology have been explained in this chapter. Therefore, this chapter gives a brief discussion about all the factors that were involved in bring the genetics in this world. With that it also explains the methods that were used for studying about the genes.

2.1. INTRODUCTION

It has been noticed that the availability of the genome sequence for the major crop plants have helped in opening-up some new possibilities for the genotyping and for the phenotyping in order to improve the crop. While, there are some more powerful statistical methods that have been developed for identifying the underlying genes of the quantitative traits.

There has been successful use of genomic predictions in the animal breeding and is now mostly used in the plant breeding method. It is being observed that the biometric statistics also supported the gene discovery when the genome wide maker gets combine with phenotyping which in studied in the large breeding collection or nurseries.

Further, the next generation sequencing and the site directed mutagenesis is being allowed for some of the original ideas that are being explored by the biotechnology in the crop plant that is to be revisited and is considered to be more precise solution to be perused.

It has been found to be desire to combine the genetics and the knowledge of plant nutrition. But it is necessary to have a precise phenotyping that is often required by a large number of a plants those are from the different environments and their growth season is still represented the major challenges that are occurring in the improvement of the nutrient use efficiency.

Later with the introduction of a DNA sequencing done in the early 1980s, for the genetic transformation of the important crop species, there is a development of a polymerase chain reaction (PCR) methods which have been used for the market assisted selection and for the next generation sequencing that has basically allowed for the cost effective development of the markers basically for the orphan crops.

Wheat is considered as one of the crops that is feeding the world and in addition to this the grain yield, quality traits for making the products have always been a target for the breeding. When germination of a mature seed take place before the harvest, which can because of some bad weather conditions, this will give the poor-quality result and is also referred as a pre-harvest sprouting (PHS).

Resistance to PHS is being controlled by many of the genes and their interactions are with the environment which is being revealed in a comprehensive review. Seed dormancy is considered as the major genetic factor that is used for controlling PHS resistance and is being controlled by the QTLs which are located on all the 21 chromosomes of a hexaploidy wheat.

There is a review on the role of the flavonoids, gibberellin signal pathways, alpha amylase, and the plant hormone abscisic acid. It is being argued that the considerable research which is still needed to be done, however, there are eight genes that have been identified by the transcriptomics, map based cloning, and comparative genomics.

It is also observed that there is a requirement for a digital and biotechnological revolution in the plant breeding which is important to develop the climate smart crops. By surveying the literature specifically on the genetic tools that is developed in order to support the crop improvement since 2000.

It is very interesting that the mutation might have been used continuously over the past few years and bibliometric searches have also highlighted some of the key papers that are basically based on the citations that can be of interest.

There has been genomic selection that is used for improving the wheat and the barley which is now routinely used by breeding the companies alongside conventional strategies. It is considered that these cereals are both the bred as spring type and winter type, which basically based on the requiring vernalization, and in case of barley it is 2 row and 6 row spike types.

Likewise, these methods are used in order to develop a plant to gain good quality product. Nowadays, mostly crops are being genetically modified for achieving a good quality product.

2.2. PRINCIPLE OF MOLECULAR GENETIC MAPPING

Molecular markers are basically considered as a DNA sequences (by using various molecular techniques) that have been used as a genetic marker that are used for identifying the individual, market traits and species association.

Therefore, these markers are said to be similar to the genes that can be mapped on the chromosome by using the principle of genetics which involves the law of independent assortment and linkage analysis. According to the Mendel's second law of independent assortment, there is either allele of a gene or molecular marker that has an equal chance to go with the allele of another gene or marker that is located on the different chromosome during the formation of a gametes.

Therefore, the recombinant and the parental type of gametes are basically formed in the equal frequencies that means that the recombination frequency is up to 50%. It is also believed that it holds the true for the genes or markers that are residing apart on the same chromosome. But this is not same for those genes that are found to be closer together on the same chromosome.

It often leads towards the reduction frequency of the recombinant type gametes which is less than 50%. This is because the chance of crossing over in every cell which is involved in the formation of the gametes is reduced.

The result of transmission of closely or tightly linked genes that are from the parents to the progeny in the higher frequency than those genes which are basically located far apart. Therefore, the genotypic frequency of the recombinant and of the parental genotypes are basically derived from the expected frequency in the given mapping population.

The recombination frequency is found to be between the two genes that are directly related to the distance that is between the two genes or markers and are being determined in term of centimorgan (1cM = 1% recombinants).

2.3. THE GOALS OF CROP BREEDING

The crop breeders are using various technologies and methodologies in order to achieve the targeted and direction changes in the nature of a plant. The advances in the science and technology are considered as a new tool that are often being developed while the old ones are the refined that are used by the breeders.

Before initiating any of the breeding project, it is important to define the clear breeding objectives that are based on the factor like consumer preference, producer needs and the needs with the environmental impact. Breeders basically aims on making the crop producers to do their job easier and even more effective which is in various ways.

They may even modify the plant structures; therefore, it will also resist the lodging and can even facilitate the mechanical harvesting. It is observed that they may also develop the plant that resist pests, because of that the farmers do not need to apply the pesticides or even apply the small amount of these chemicals.

If there is no application of pesticides in the crop production than that means that there is less environmental pollution from the agricultural sources. It is believed that the Breeders may also develop high yielding varieties, so that the farmers can also produce more for the marketing purpose in order to meet the demand of the customers while improving their own income.

The term cultivar is being reserved for the variants those are deliberately created by the plant breeders. When breeders think of the consumers, they also think for developing food with the higher in nutritional values and more flavorful product.

From the term higher nutritional value, it simply means reduction of illness in the society like blindness, rickettsia, which may be caused by the consumption of the nutrient – deficient foods. In many of the developing regions there are staple foods like rice and cassava, that often lack some certain essential amino acids or the nutrients.

Plant breeder are also believed to target the traits of industrial value. For example, fiber characteristics of fiber crops like cotton can be improved. While the oil crops can be also improved in order to yield the high amount of specific fatty acids, for example, the high oleic content sunflower seed.

According to the latest advances in the technology, specific genetic engineering technologies are often being applied in order to enable the plants that is to be used as a bioreactor to produce some certain pharmaceuticals which is called as biopharming or simply pharming.

It is believed that the technological capabilities and the needs of the society which is in the past restricted by the plant breeders in order to achieve the modest objectives. For example, adaption to the production environment, and product appeal.

2.4. THE CONCEPT OF GENETIC MANIPULATION OF PLANT ATTRIBUTES

The work of Gregor Mendel and further advances in the science that are followed by the Mendel discoveries often established that the plant traits are being controlled by the hereditary factors or genes that are consisting of DNA.

These genes are being expressed in an environment for producing a trait. In order to change the trait or the expression, it is important for the one to change the nature or genotype, and to modify the nature which is being expressed.

Changes in the essential of the environment which entails modification of the growing or production conditions. This may be further achieved through the agronomic approach. For example, the application of the production inputs like irrigation and fertilizers.

While the approach is found to be effective in enhancing some of the certain traits, with the fact that these supplemental environmental factors are being removed and the expression of the plant traits are reverting towards the status quo.

Whereas, on the other hand the plant breeder is seeking some help to modify the plants with respect to the expression of a certain selected attributes that is by modifying the genotypes.

2.5. GENETICALLY MODIFIED PLANTS

Genetically modified technologies are being highlighted because of the most significant and recent developments that took place in the plant breeding. GM technologies are allowing the plant genomes to be enriched with the DNA of any organism, and that would lead towards introduction of a new variation into a gene pool.

It is considered that transgenic technologies have become an integral part of the plant breeding technique, it is also important to define the traits that are considered to be targeted for the purpose of introducing and the time that is required to develop a transgenic product.

It is considered that typically, it requires ten generations for creating a transgenic variety that can be used for testing, and which can be similar to the time frame for testing and for developing a line through the exotic breeding methods.

It is also considered to be important to note that once there is an exotic library that has been generated, and the time that is required for its use in the breeding for other traits is considerably short.

Figure 2.1: GM cotton crop.

Source: Image by Pixabay

There was an identification of a natural insect toxin that is present in the bacterium *Bacillus thuringiensis* which made it possible to introduce a new insect that is of tolerance trait into a crop plant. This introduction has helped significantly by reducing the amount of insecticides that are often released into the environment.

There was introduction of resistance towards the herbicide glyphosate from the plants of *Agrobacterium tumefaciens* which is a bacterium that mainly causes crown gall disease in some of the plants, which often provides an efficient method of controlling the weed and therefore increased the productivity.

These GM plants are not most widely used for cultivation, and the expectation from this is that the rate at which the unique traits are being introduced, are considered to benefit the end consumers, and will also increase in the next century.

One of the first GM variety that enters the market was the Flavr Savr tomato, Which was considered to be codified in order to slower down the soften rate than the normal tomatoes, which basically allowed it to be picked at a large stage of maturity when it is considered that the flavor of the tomato is more developed.

Flavr Savr tomato have been created by using the antisense technology that often inhibits the expression of the enzyme poly-galacturonase, which is observed to be associate with the cell wall softening. At the same time, it is being observed that the tomato varieties that are heterozygous for the naturally occurring, single mutation gene RIN (Ripening inhibitor) which took the tomato market by storm and is still dominating it. Whereas, it is observed that the Flavr Savr has a little no market share.

This basically happens because the effect of RIN on the extending shelf life is much significant. Therefore, like wise examples are indicating that there are some of the variations that are being introduced into the plants by the transgeneses which is used for suppressing some of the certain gene products that occurs naturally or is being induced in mutants.

It can be exploited in a cost-effective manner. It should be noted that the application of the GM technologies in the plant breeding should be only in cases where there are no other classical gene alternatives.

2.6. MENDEL AND PEA PLANTS

Gregor Mendel was born in the year 1922 and grew up on his parent's farm which was in Austria. Gregor Mendel was well in school and became a monk. Mendel went to the university of Vienna, where Mendel studies about the Math and Science.

Mendel's processor was the one who forced Mendel to learn about the science through the experimentations and with that to use the math logics to make sense of result. Mendel was best known for his work on pea plant.

2.6.1. Blending Theory of Inheritance

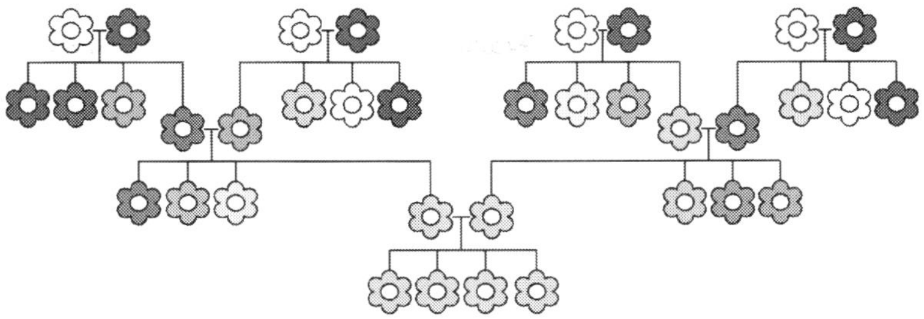

Figure 2.2: Representation of blending theory of inheritance.

Source: Image by Wikimedia Commons

Before the arrival of Mendel theory, the blending theory of inheritance was popular. This theory often explains that the offspring is a blend or mix of the characteristics of their parents. Then, Mendel noticed that the plants present in the garden were not a blend of the parents.

For example, a short plant and a tall plant are offering that either the new generation will be tall or will be short but they cannot be of medium height. This as a result, led Mendel to question the blending theory. Mendel wondered that there might be some principle that can well explain about the characteristics that are inherited.

Therefore, Mendel decides to experiment with the pea plant to find out the real reason behind the inheritance. Mendel in the experiment, decided to work on almost 30,000 pea plant for the next several years.

2.6.2. Reason behind Choosing Pea Plant

The reason behind choosing pea plant (*Pisum sativum*) by Mendel is that Pea plant are considered that they comparatively faster and are also easy to raise. Pea plants are also having several visible characteristics that may vary.

These characteristics includes pod color and form, seed color and form, flowers on stems and stem length, flower color and placement of pods. Each of these characteristics are considered to have a common trait. For example, seed form can be round or wrinkled and flower color can be white or purple or violet.

2.6.3. Controlling Pollination

In order to research about the reason behind how the characteristic are passing from parents to offspring, Mendel need to control the pollination. Pollination is considered as a fertilization step that is occurring in the sexual reproduction of the plants.

Pollens are consisting of tiny grains that are representing the male sex cell or gametes of the plants. These grains are produced by a male flower part which is called as the anther. The pollination basically occurs when the pollen is transferred from the anther to the stigma of the same flower or can be of another flower.

The stigma is considered as a female part of a flower. Stigma often helps in transferring or passing the pollen grains to the female gametes present inside the ovary.

Pea plants are considered to be naturally self-pollinating plants. In the self-pollination method, the pollen grains from the anther on one plant are being transferred to the stigma of the flowers on the same plant. Mendel was seen interested in the offspring of two different parent plants, so Mendel had to prevent the self-pollination.

Mendel removed the anthers from the flowers of some plants in the experiment. Then Mendel pollinated them by using hands with the pollen from the other parent plants which was Mendel choice. When the pollen from one of the plants is being fertilized with another plant of the same species, this process is called as a cross-pollination.

The offspring that is achieved as a result from such cross are called as hybrids. The term hybrid refers to any of the offspring that is a result from the breeding of two genetically distinct individuals.

2.6.4. Mendel's First Set of Experiments

First, Mendel experiments by using one characteristic at a time. Firstly, Mendel began with the flower color. Mendel cross pollinated the violet and white flowered parent plants. In the experiment, the parent plant is being referred to as P (for parents) generation.

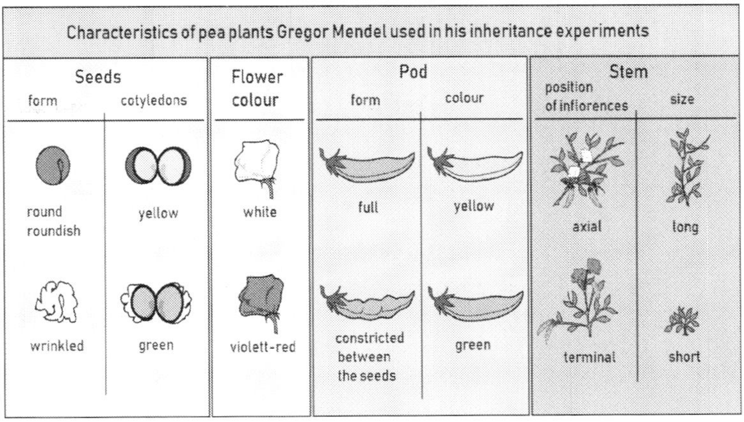

Figure 2.3: Characteristics chosen by Mendel for the study of inheritance.

Source: Image by Wikimedia Commons

F_1 and F_2 Generations

F_1 generation is basically the offspring from the P generation. In the below it can be clearly seen that all the plants in the F_1 generation are containing violet color flowers only. Where none of the plant is having white flower.

After this, Mendel wondered about this and started thinking about the white flower characteristics. Mendel further assumed that there is some type of inherited factors that are producing white flowers and some other inherited factors are often producing violet flower.

So, did the white flower factors got disappear in the F_1 generation? If this is true, then the offspring of the F_1 generation - which is called as F_2 generation should have all the violet flowers like their parents. So, in order to test this prediction, Mendel again allowed the F_1 generation plants to self-pollinate.

Mendel got surprised by the result. Mendel noticed that some of the F_2 generation plants had white flowers. Mendel studies hundreds of F_2 generation plants, and from there Mendel noticed that for every three violet flowers plant there was an average of one white flowered plant.

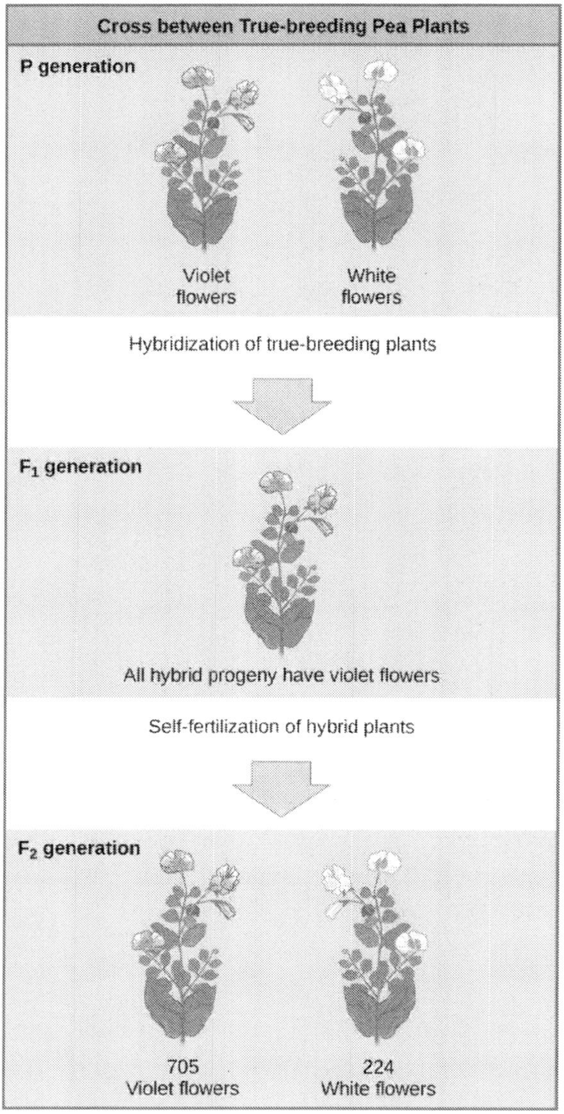

Figure 2.4: Representation of F_1 and F_2 generations.

Source: Image by Wikimedia Commons

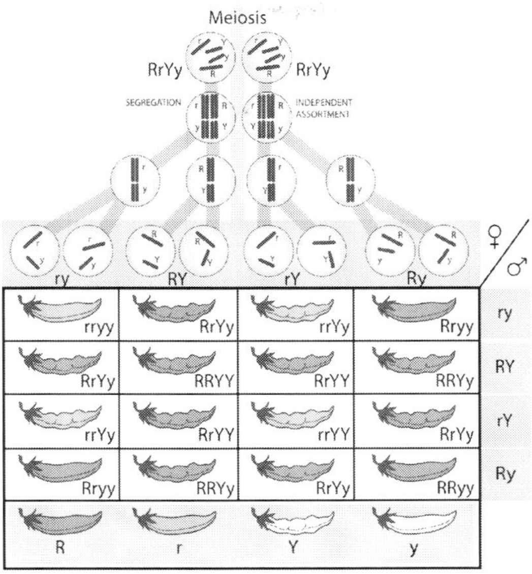

Figure 2.5: Mendel law of segregation and independent assortment.

Source: Image by IUPAC at

Law of Segregation

After then, Mendel did the same experiment for all the seven characteristics. In each of the seven cases, one value of the characteristic seems to disappear in the F_1 plants and then it is being observed in the F_2 plants. In each of the cases, there are about 75 percent of the F_2 plants that had one value of the characteristic and 25 percent are having other values.

Based on these observations, Mendel formulated the first law of inheritance. This law is called as the law of segregation. This law states that there are two factors that are controlling given set of characteristics, where one of which are dominating the other and with that these factors are separate and during their parent reproduce, they go to different gametes.

2.6.5. Mendel's Second Set of Experiments

Mendel in the second set of experiment wondered that whether there are different characteristics that are inherited together. For example, is it true that purple flowers and tall stems are always inherited together? Or they do these two characteristics which show up in the different combinations in offspring?

In order to answer for these questions, Mendel next investigated two characteristics at a time. For that Mendel crossed plants with the yellow round seeds and plants with green wrinkled seeds.

F_1 and F_2 Generations

In the experiment, Mendel observed that the plants those are in F_1 generation were all the plants are alike. All the plants had a yellow round seed like one of their two parents. Later when the F_1 generation is self-pollinated then their offspring in the F_2 generation were showing all the possible combination of the two characteristics.

In the F_2 generation, some of the offspring were having green round seed, for example, some of them were having yellow wrinkled seeds. These combinations were the characteristics that was not present in the parent plants in F1 generation or P generation.

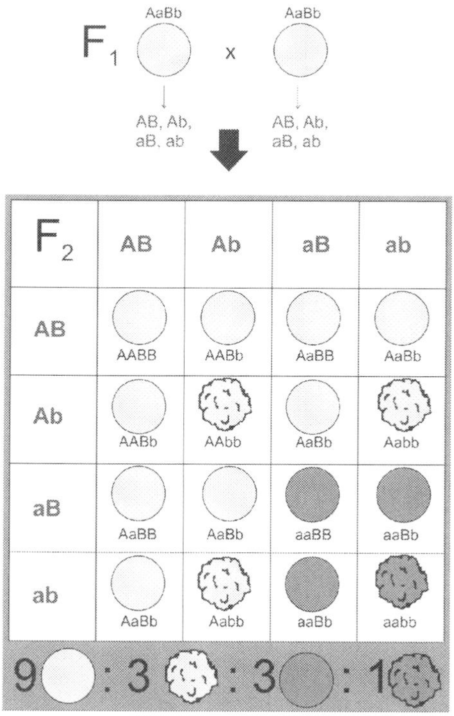

Figure 2.6: Representation of second experiment of Mendel.

Source: Image by Wikimedia Commons

Law of Independent Assortment

Mendel further repeated this experiment by taking other combinations of characteristics. Some of the characteristic combination were like flower color and stem length. Each time the result was kind of similar as in the above combination.

This experiment results into Mendel's second law and this law was known as a law of independent assortment. This law states that the factors that are controlling these different characteristics are being inherited independently of each other.

2.7. TECHNOLOGIES TO INTRODUCE NEW GENETIC VARIATION

There are several genetic technologies that have been developed in order to expand the range of the genetic variations that are basically beyond that is found in the allelic variations of the native genes in the diverse germplasm.

The mutagenic effects of the x-ray, certain chemical discoveries, and ionization radiations that have been discovered in the early to mid-twentieth century that have led to the use of a mutagenesis which is used as a tool in the plant breeding programs.

Most of the mutation process are tend to have a negative consequence from the perspective of crop improvement. But it is considered that the dedicated efforts done in the mutation breeding have led to the release of about hundreds of improved cultivars (Maluszynski, 2001).

It should be noticed that the semi dwarf stature has been developed by the process of mutation and is being utilized in a number of crops, which includes rice, barley, wheat, and the sunflower. These crops are utilized because to prevent the plants from falling over prior to harvest, which is considered as a condition that can result in the gain in yield loss (Maluszynski, 2001; Rutger, 1992).

Recently, the process of Tilling (is used for targeting the induced local lesions in the genomes) has been developed in order to identify the induced point mutations that takes place in the specific genes of interest in the population of individuals those are chemically mutagenized by the traditional means (Till et al., 2003).

High throughput Tilling is being allowed for the rapid and for the low-cost recovery of the plants with induced point mutations in the key genes. Tilling further facilitates the evaluation of the diverse plant germplasm resources for the mutant alleles that may be influencing the trait of performance.

Late in the 20th century, the development of the genetic transformation techniques has been providing an opportunity for the plant breeders to introduce the transgenes into the crop species. Therefore, there is greatly expanding of the gene pool that is otherwise seems to be limited to the species that are cross-pollinate with the crops.

Figure 2.7: Representing some of the example of genetically modified crops.

Source: Image by Peakpx

Transgenics have been revolutionized for the plant breeding in the several major crops which includes soybeans, canola, maize, and cotton. This is done by facilitating the access towards the genetic variation which is done by contributing towards the pest, disease and weed management that is not present in the native gene pools.

Transgenes are often conferring the improved nutritional qualities, abiotic stress resistances, and other key traits that are being developed. Currently there is use and benefits of the transgenes, but they are limited or non-existent in some of the crops because of the regulatory policy and consumer acceptance.

2.7.1. Doubled Haploidy

For many of the crops, the evaluation has been done by using the homozygous lines that are developed by the several generations of the self-pollination. Such homozygosity is "fixing" the genotypes in order to stabilize the trait expressions, which is allowing a line to breed true.

Developing a homozygous line from a cross that is between the two different parental lines are however said to require six to eight generation of selfing, which might can take a long time. Doubled haploid are offering a fastest means for developing a homozygous line.

There is a possibility that haploid plants may occur spontaneously in the nature or can be through various innovative methods, by using either paternal or may be maternal gametes. There are further innovations that can enable the breeders to double the genome of a haploid which can takes place via in vitro or in vivo means.

This is often resulting in the homozygous lines with sufficient seed used for testing in a matter of two to three generations. This technology has been implemented on routinely basis in wheat and corn. There are many commercial corn breeding programs that might have shifted towards nearly 100% of the doubled haploids.

Therefore, the reduced development time often results in cost saving, which is more precise field of testing, and is faster time to market with the improved cultivars.

2.7.2. Precision Phenotyping

Evaluation of the plant performance often becomes critical aspect of plant breeding. Phenotyping basically involves the assessment of performance for the traits that are associated with the breeding targets. There are some new and innovative approaches and technologies that have been developed in order to facilitate these evaluations.

It is considered that these approaches do not take the place of yield performance while testing in order to finalize the decision taken about the release of new cultivars. Therefore, the novel approaches towards the phenotyping often enables the collection of the additional information that basically serves to enhance and also to make better selection decision.

It is believed that the primary features of a new phenotyping technologies basically involve the automation which often enables more measurements through the time or faster data collection. It also focuses on the secondary traits that are considered to be strongly associated with the target traits that are yet easier or considered to be less expensive in order to measure, and collection of the information which can facilitate the insight into the environmental condition which are impacting the phenotype.

Particularly, the cost and the weight of the cameras has come down to an extend at which there are several cameras that can be attached to the drone or unmanned aerial vehicles and they can even take a large number of images of a developing crop.

Figure 2.8: Image representing different characteristic of a same plant.

Source: Image by An Endless Falling – WorldPress.com

The above image is representing the number of parameters that can be easily measured. By using the normal visible light cameras, the growth rate of a crop can be easy measured and can be easily plotted too (Grieder, Andreas, and Achim 2015).

The above image is also allowing to detect the abnormal color of the crops that might be due to factors like disease outbreak or nitrogen deficiency. There can be use of long wave infrared or thermal cameras that can help in determining that if plants are under some kind of stress or they are suffering from some disease.

For example, under drought stress plants, they are considered to be less able to cool themselves which can be via transpiration and it can tend to show an increase in the canopy temperature. It is now possible to use the visible, hyperspectral, and near infrared cameras that will help in capturing hundreds of images which is over the whole electromagnetic spectrum (Araus and Cairns, 2014; Fahlgren, Gehan, and Baxter, 2015).

There is a ground-based sensor that can be used for measuring the moisture availability in the soil (Araus and Cairns, 2014), which also assess the canopy temperatures in order to help to identify the drought tolerant genotype (Andrade-Sanchez et al., 2013) or can even predict the traits like biomass yield (Pittman et al., 2015).

It is believed that all these data can be fed into a model of the crop performance. In order to improve the plants, the breeders are selected for the traits which is required, to measure the yield, measure proteins, to get high yielding plants, for the plants which are high in protein and so on.

Directly measuring of the phenotype can be considered to be time consuming, difficult or can be feasible for some traits, even with the advanced machinery. Otherwise, the breeders can also practice "indirect" trait selection. Therefore, they can also measure another trait that are correlated with the traits of interest that is evenly easily and accurately measured.

Some of the techniques like near infrared reflectance spectroscopy (NIRS) which have been used for a decade for predicting the parameters such as forage or gain composition (Roberts, Workman, and Reeves, 2004) for avoiding the laborious wet laboratory or an in-vivo procedures.

Recently, on-harvester NIRS equipment (Montes, Melchinger, and Reif, 2007) which has been developed in order to measure these traits "on the fly", which is further increasing throughput by simply not requiring separate lab analyses.

There is another opportunity for selecting the indirect traits that can be determined by observing the relationship between the phenotypic traits. For example, total biomass of the plant is often being associated with the leaf area and for the duration of the vegetative growth.

It is believed that some of the associations that takes place between the traits can be quite complex. For example, canopy temperature that can be even linked to the heat tolerance because of the plant that is found to be cool is often able to transpire more water than the hot plants and this even protects it from the high temperatures.

The traits are also observed to be associated with the root depth because of the plant that can even track the water down to the soil profile which might be late in growing season which is better be able to access the water (Reynolds and Langridge, 2016).

Therefore, the temperature of the canopy can be even used for selecting the plants that might have deep roots. The selection for the drought tolerance has been considered to be taken on the greater importance with the climate change scenarios.

Even though it is considered as a very complicated trait because of the timing and severity of the drought stress that can have different effects on the yield and can also use as an economically important trait. In the maize, at the moderate drought stress is initiated near the flowering time, which helps in gaining the yield which is associated with the anthesis-silking interval (ASI).

Which is considered as a stress response indicator. ASI can be further used to screen the improved drought tolerance (Ribaut et al., 2004).

As a result, these measurements are aiming to support the selection for the traits that will be considered as important for the farmers and for the end users of the harvest crop. Breeders are often seeking tools from the precision phenotyping datasets that are

- feature assays, that are considered to be low cost and with the high throughput. So that they can screen large populations at reduced cost compared to conventional screening methods;
- facilitate high accuracy in measurement;
- correlate highly with the traits of importance to their programs, so that they can replace time consuming, unreliable, expensive measurements or difficult; and
- offering high heritability (and repeatability) and gaining rapid genetic through selection (i.e., high control of environmental variation).

2.7.3. Hybridization Systems

In the early 1761, the experimentation into the hybrid vigor or heterosis was being conducted by the Joseph G. Kölreuter, who was a German botanist (Kolreuter 1761–1766). The exploitation of this phenomenon was done to increase the grain yield of a maize which is considered to have a long history.

It was observed that the American Indian tribes often planted different types of corns which is close to the proximity to increase the yields (Obolensky, 1958). Because of the hybridization that can translate it to the yield increase of about 20-50% in the self-pollinated or in the open – pollinated crops (Tester and Langridge, 2010).

As it is true, that there are many crops that are basically grown either significantly or mostly as a hybrid like in case of tomato, rice, wheat, sunflower and canola and this kind of hybridization often plays an important role in cultivars of cotton, wheat, triticale, and barley.

Capitalization on the hybrid vigor, hybridization can be used as a straight forward way for increasing the yield per land unit. Therefore, the expression of the heterosis often requires the use of a genetically dissimilar parent lines. Therefore, it can also demand for the complementarity that takes place between the parent lines in terms of favorable alleles.

Currently, there are higher yielding hybrid crops that are accounting for a limited portion of the global cultivated area. For example, there are less than 1% of the total world wheat area which is found to be planted with the hybrids (Longin et al., 2012).

Therefore, there is an increased use of hybridization that has a tremendous potential for increasing the agricultural productivity. To use the hybridization as a mean that can boost the crop productivity, that will create a way through which the F_1 generation seed is needed, the system should be economical and practical.

If the crop species is found to be dioecious (there is present of separate male and female plants), then the seed production can be easily arranged in order to produce the F_1 farmer seed. **B**ecause both the pollen distribution and seed set can be well controlled.

There are other approaches that involves a means in order to control the pollen exposure to ensure that there is only pollen from the plant that is intended as a male which is involved in the seed production on the plants which are intended as females.

This basically includes the physical removal of a male reproductive plant parts through the mechanical means like detasseling, there is a use of cytoplasmic or nuclear genetic system in order to elicit the male sterility in the female parent plants, and chemically or transgenically induced male sterility.

It is required to find out the means to transfer the pollen from the desired male parent to the desired female parent plants. There is an innovative technology that have been developed in order to accomplish these objectives and even facilitate the hybrid seed production that is economically and is in the large scale.

In case of U.S. hybrid maize production fields, the physical detasseling has been considered as a predominant method that is used to control the pollination (Wu et al., 2015). Despite of the fact that the removal of a vegetative material it can even decrease yield of the seed up to 40% (Wych, 1988).

The Cytoplasmic Male Sterility (CMS) systems are considered to be employed, but they are not effective as in the maize germplasm and it can even fail to maintain the sterility under the environmental stress conditions (Wu et al., 2015).

Cytoplasmic male sterility is considered to be very reliable and is often used to produce the new cultivars of several horticultural crops. These horticultural crops include all the hybrid cultivars of the table beets, onions, petunias, and carrots as well as some of the hybrid cultivars of sugar beet as well as some of the hybrids of canola and alfalfa (Havey, 2004).

2.8. QUANTITATIVE GENETICS

There are some population genetics and some quantitative genetics that are found to be closely related fields, both of them are found to deal with the genetic basis of phenotypic variation that takes place among the individuals in a population.

Population genetics is traditionally focusing on the frequency of the alleles and genotypes. Whereas, the quantitative genetics is focusing on the linking of the phenotypic variations of the complex traits towards its underlying genetic basis which basically enables the researchers to better understand and even predict the genetic architecture and a long term changes in the populations.

These changes are done in order to predict the response towards the selection of the given data on the phenotype and relationships of a individuals in the population. Historically it is believed that the quantitative genetics has its roots in the statistical abstractions of the genetic effects that is being firstly described by the Karl Pearson and Ronald Fisher in the early 1900s.

Therefore, forgoing is often representing the classical view of the quantitative genetics. Accordingly, the molecular view of the quantitative genetics is focusing on the use of the molecular genetic tool like bioinformatics, computational biology and genomics which help in revealing the link between the gene and the complex phenotypes (quantitative traits).

Genes that are often controlling the quantitative traits are basically called as quantitative trait loci (QTLs). The molecular based QTL analyses are being used for evaluating the coupling associations of the polymorphic DNA sites with the specific phenotypic variation of the quantitative and complex traits and with that also analyzing their genetic architectures. There is an evidence of a paradigm shift that takes place in the field of quantitative genetics.

2.8.1. Classical Quantitative Genetics

- **Quantitative Trait**: It is being observed that most of the traits that are encountered in the plant breeding are mostly quantitatively inherited. There are many genes that are controlling such traits, where each of them is contributing a small effect to the overall phenotypic expression of the traits.

The variation in the quantitative trait is considered as an expression which is without the natural discontinuities. The traits those are exhibiting continuous variation are also known as a metric trait. If there is any attempt that is used to classify such traits into a distinct group then that is only possible through arbitrary.

For example, height is considered as a quantitative trait. If the plants are being grouped into all tall versus short plants, then their relatively tall plants can be only found in the short group and likewise, if short plant is to be find then this can be possible in the tall group.

- **Qualitative Genetics Versus Quantitative Genetics:** There are some major ways through which quantitative genetics and the qualitative genetics differ easily and this difference is summarized as following:
- **Nature of Traits:** The qualitative genetics is being concerned with the traits that have Mendelian inheritance and it can also be described according to the kind and also ambiguously categorized.

These quantitative traits are being described in terms of degree of expression of the traits that is rather than the kind.

- **Scale of Variability:** Qualitative genetics are basically the traits that is providing discrete or discontinuous phenotypic variations. Quantitative genetic traits are those traits which are producing the phenotypic variation that basically spans the full spectrum.
- **Number of Genes:** In the qualitative genetics, it is found that the effects of a single genes are readily to be detectable. Whereas, in the quantitative genetics the single gene effects are not properly visible. Rather than that, the traits are under the polygenic control which means that the genes are with small indistinguishable effects.
- **Mating Pattern:**the qualitative genetics is being concerned with individual mating and their progenies. The quantitative genetics is basically concerning with a population of a individuals that may even comprise of a diversity of mating kinds.
- **Statistical Analysis:** The qualitative genetics analysis is considered to quite straight forward which is basically based on the count and ratios. Whereas, on the other hand the quantitative analysis is provided estimate of the population parameters which means attributes of the population from which the samples are being obtained).
- **The Environment and Quantitative Variation:** It is believed that all the genes are expressed in an environment where phenotype is approximately ¼ genotype þ environment effect. Therefore, the quantitative traits are tending to get influence to a greater degree than the qualitative traits.

It is being observed that under significantly large environmental effects, the qualitative traits which are basically controlled by one or few of the major genes are consider to exhibit the quantitative trait inheritance pattern. There is a strong environmental influence that are causing the otherwise distinct classes to overlap.

- **Polygenes and Polygenic Inheritance:** The quantitative traits are being controlled by multiple genes or by the polygenes. Here polygenes are basically the gene with the effects that are too small to be individually distinguish.

They are even sometimes called as a minor gene. In the polygenic inheritance, the segregation occurs at a very large number of loci that is affecting the traits. It is believed that the phenotypic expression of the polygenic traits which is basically susceptible towards significant modification which is by the variation in the environmental factors to which the plants in the population are being subjected.

It is observed that the polygenic variation cannot be classified into the discrete groups, which means there is variation continuously. This is basically because of a large number of segregation loci, where each of them is with small effects that is not even possible to identify the individual gene that is affecting in the segregating population or which is meaningfully describing the individual genotype.

Instead of this, there is a use of biometric which is used to describe the population basically in terms of means and variances. Continuous variation is cause by the environmental variations and the genetic variations that are due to the simultaneous segregation of many of the genes that are affecting the trait.

Later, these effects are being converted in the intrinsically discrete variation towards a continuous one. Biometric genetics is often used to distinguish between the two factors that are causing continuous variability to occur.

According to another aspect of the polygenic inheritance which is different combination of the polygenes that can produce a phenotypic expression. Therefore, it is found to be difficult to measure the role of the environment on the traits expressions because it is found to be very difficult to measure the environmental effects which is one the plant basis.

So, it is important for a breeder who is attempting to breed a polygenic trait that to evaluate the cultivars in an environment that is basically similar to the prevailing in the production region. It is beneficial for plant breeding if there is a tight linkage of the polygenes that is called as polygenic block or a linkage block that is having a favorable effect on the traits of interest towards the breeders being discovered.

2.8.2. Number of Genes Controlling a Quantitative Trait

Polygenic inheritance is being characterized by the segregation process which takes place at a large number of loci and is affecting the traits also. The biometrical procedures have been often proposed in order to estimate the number of genes that are involved in a quantitative trait expression.

Therefore, such kind of estimates, apart from not being reliable, it is considered that it has limited practical use. It is believed that genes may also differ in the magnitude of their effects on the traits that do not mention that there is a possibility that the modifying gene may also effects some of the certain genes.

- **Modifying Genes**: It is believed that one of gene may have a major effect on one trait while the other one gene may have some minor effect on another trait. There are basically many gene in plants that may without any known effects. Besides the fact that they are modifying the expression of a major gene which is either by enhancing or diminishing it.

It is observed that the modifier genes may be subtle, which means that there are some slight variations in the traits that are like shape and shades of color of flowers or in fruits, and variation in the taste and aroma.

Those traits are being modifies which are basically concern to the plant breeder as they are conducting the breeding programs for improving the quantitative traits that is involving many of the major traits that are of interest.

2.9. CYTOGENETICS

Cytogenetics is basically a study of the structure and properties of the chromosomes, development through mitosis, chromosomal behavior during the somatic cell division in growth, chromosomal influence on the phenotype, the factors that are causing the chromosomal changes (Hare & Singh, 1979), and germ cell division process in the reproduction through meiosis.

There is a discovery of new technique that is often used to see the improvement in the existing techniques or the new combinations of a well-established techniques that are being followed by the progress in the bioscience.

This is considered as a strikingly exemplified which is by the development of the cytogenetics in the last 100 years.

2.9.1. Use of Cytological Approaches to Study the Behavior of Plant Meiotic Chromosomes

The study of plant meiosis often revealed the value of biochemical approaches and combining genetics with the cytology in order to visualize and even characterize genome of an organism.

It is being observed that the cytology and behavior of a meiotic chromosomes that are in the plants have been dissected for more than a century by a combination of a cytological, biochemical, and genetic approaches (Wilson, 1925 ; McClintock, 1929a, b ; Darlington, 1932 ; Stern and Hotta, 1974 ; Golubovskaya, 1979 ; Golubovskaya and Khristolyubova, 1985 ; Ross et al., 1996 ; Dawe, 1998 ; Armstrong and Jones, 2003 ; de Jong, 2003 ; Jenkins et al., 2005 ; Ma, 2005 ; Pawlowski and Cande, 2005 ; Hamant et al., 2006 ; Jiang and Gill, 2006).

Recently, the live-cell imaging with the cultured anthers has been in a use to elucidate the cell biology in the higher plant of meiosis (Chan and Cande, 2000; Shaw, 2006; Sheehan and Pawlowski, 2009). That cytological technique that was facilitating direct visualization of a meiotic chromosomes which is in situ is considered to be traditionally relied on a combination of tools.

This tool is basically including bright field microscopy, electron microscopy (EM), fluorescence in situ hybridization (FISH), and the protein immunolocalization in order to probe the architecture and behavior of the meiotic chromosomes in the animals and plants.

Recently, there is new ultrahigh resolution of the light microscopic techniques like structured illumination (SI) microscopy, which have been used to obtain the exceptional images of the chromatin and chromosomes structure which is during the meiosis and takes place in a variety of organisms that are including plants also. (Carlton, 2008 ; Gustafsson et al., 2008 ; Shao et al., 2008).

2.10. CONCLUSION

Plant breeding is basically an effort made by the humans to improve the nature by maintaining the heredity of the plant. The changes that are made in the plants are permanent and are heritable too. The professionals who are conducting this process are called as plant breeders.

Plant breeding technique is basically done to improve the existing characteristics of a plant and to develop new characteristic that will be beneficial for the plants and their end users. Earlier the plant was considered to provide the result according to the situations.

As these new technologies have developed, this helped the plant to give good result in any of the situation and protect the plant from getting ruin.

Therefore, for protecting and for good quality product , several methods have been used.

Science and technology have played their important role in developing the new verities that can benefit the society. Therefore, there have been several techniques and methods came into implementation that helped in satisfying all the needs.

For this purpose, there have been several experiments that took place to understand the plant and their fertilization process. Likewise, Mendel gave two laws to explain the inheritance in the plants. For generating new species several methods like hybridization, cloning, and many other methods were included to explain, develop new varieties and their roles.

Some experiments were done to check the quality of the plant and some to check the quantity of the plant. Nowadays, genes have been used for modification which means developing new changes in the offspring, hybridization which means developing offspring with the good or desired quality of a plant, cloning which means developing the same variety or copy of the parent plant.

In this chapter there is a brief discussion about all the steps that were taken to bring the changes in the plant quality, variety, and quantity and with that it also explains the different steps that are involved.

REFERENCES

1. Acquaah, G., 2012. *Principles of Plant Genetics and Breeding.* [ebook] A John Wiley & Sons, Ltd, Publications. Available at: <https://gtu.ge/Agro-Lib/Principles%20of%20Plant%20Genetics%20and%20Breeding.pdf> [Accessed 4 June 2020].
2. Baenziger, P., Mumm, R., Bernardo, R., Brummer, E., Langridge, P., Simon, P. and Smith, S., 2017. *Plant Breeding and Genetics.* 57th ed. [ebook] Available at: <http://pba.ucdavis.edu/files/259473.pdf> [Accessed 4 June 2020].
3. Biology LibreTexts. 2019. *Mendel'S Laws of Heredity.* [online] Available at: <https://bio.libretexts.org/Bookshelves/Introductory_and_General_Biology/Book%3A_General_Biology_(Boundless)/12%3A_Mendel's_Experiments_and_Heredity/12.3%3A_Laws_of_Inheritance/12.3A%3A_Mendel%E2%80%99s_Laws_of_Heredity> [Accessed 4 June 2020].
4. Choudhary, A. and A G, V., 2012. *Breeding: A Perspective.* [ebook] Available at: <https://www.researchgate.net/publication/280714722_Glossary_of_Plant_Breeding_A_perspective> [Accessed 4 June 2020].
5. Crusio, W., 2020. *An Introduction to Quantitative Genetics.* 2nd ed. [ebook] CRC Press. Available at: <https://www.researchgate.net/publication/28762168_An_introduction_to_quantitative_genetics> [Accessed 4 June 2020].
6. Daniell, H., 2002. *Transgene Containment Via Maternal Inheritance and Male Sterility in GM Crops.* [ebook] Available at: <https://www.researchgate.net/publication/239543356_Transgene_containment_via_maternal_inheritance_and_male_sterility_in_GM_crops> [Accessed 4 June 2020].
7. Higgins, N., 2001. Chromosome Structure. *Encyclopedia of Life Sciences*, [online] Available at: <https://www.researchgate.net/publication/227550777_Chromosome_Structure> [Accessed 4 June 2020].
8. Kumar, J., Pratap, A. and Solanki, R., 2009. BASICS OF MOLECULAR GENETIC MAPPING AND QTL ANALYSIS IN PLANTS. *Advances in Biotechnology*, [online] pp.35-52. Available at: <https://www.researchgate.net/publication/287573445_Basics_of_Molecular_Genetic_Mapping_and_QTL_Analysis_in_Plants> [Accessed 4 June 2020].

9. Murphy, S. and Bass, H., 2011. Genetics and Cytology of Meiotic Chromosome Behavior in Plants. *Plant Cytogenetics*, [online] pp.193-229. Available at: <https://www.researchgate.net/publication/236345887_Plant_Cytogenetics> [Accessed 4 June 2020].
10. Novak, F. and Brunner, H., 1992. *Plant Breeding: Induced Mutation Technology for Crop Improvement.* [ebook] IAEA BULLETIN. Available at: <https://www.iaea.org/sites/default/files/34405682533.pdf> [Accessed 4 June 2020].
11. Rasmussen, S., 2020. *Https://Www.Mdpi.Com/2073-4395/10/3/439/Pdf.* [ebook] Available at: <https://www.mdpi.com/2073-4395/10/3/439/pdf> [Accessed 4 June 2020].
12. Simpson, A., 2001. *Improving Plant Breeding with Exotic Genetic Libraries*. [ebook] Available at: <https://www.supagro.fr/ress-tice/ue1-ue2_auto/Etude%20cas%20tomate/Zamir2001_SAM.pdf> [Accessed 4 June 2020].
13. Singh, B., Singhal, N. and Maurya, N., 2001. *GENETICS AND PLANT BREEDING AND SEED SCIENCE AND TECHNOLOGY.* [ebook] New Delhi: Shri Arvind Chakravarty. Available at: <https://icar.org.in/files/mGen.pdf> [Accessed 4 June 2020].
14. Thirumulu, K., 2011. Cytogenetic Techniques in Diagnosing Genetic Disorders. *Advances in the Study of Genetic Disorders*, [online] Available at: <https://www.researchgate.net/publication/221920029_Cytogenetic_Techniques_in_Diagnosing_Genetic_Disorders> [Accessed 4 June 2020].
15. Wakim, S. and Grewal, M., 2020. *8.2: Mendel's Experiments and Laws of Inheritance*. [online] Biology LibreTexts. Available at: <https://bio.libretexts.org/Bookshelves/Human_Biology/Book%3A_Human_Biology_(Wakim_and_Grewal)/08%3A_Inheritance/8.2%3A_Mendel's_Experiments_and_Laws_of_Inheritance> [Accessed 4 June 2020].
16. Andrade-Sanchez et al., 2013
17. Araus and Cairns, 2014
18. Armstrong and Jones, 2003
19. Carlton, 2008
20. Chan and Cande, 2000
21. Darlington, 1932
22. Dawe, 1998

23. de Jong, 2003
24. Fahlgren, Gehan, and Baxter, 2015
25. Golubovskaya and Khristolyubova, 1985
26. Golubovskaya, 1979
27. Grieder, Andreas, and Achim, 2015
28. Gustafsson et al., 2008
29. Hamant et al., 2006
30. Havey, 2004
31. Jenkins et al., 2005
32. Jiang and Gill, 2006
33. Joseph G. Kölreute, 1761
34. Longin et al., 2012
35. Ma, 2005
36. Maluszynski, 2001
37. McClintock, 1929a, b
38. Montes, Melchinger, and Reif, 2007
39. Obolensky, 1958
40. Pawlowski and Cande, 2005
41. Pittman et al., 2015
42. Reynolds and Langridge, 2016
43. Ribaut et al., 2004
44. Roberts, Workman, and Reeves, 2004
45. Ross et al., 1996
46. Rutger, 1992
47. Shao et al., 2008
48. Shaw, 2006
49. Sheehan and Pawlowski, 2009
50. Stern and Hotta, 1974
51. Tester and Langridge, 2010
52. Till et al., 2003
53. Wilson, 1925
54. Wych, 1988

3

Crop Breeding Objective and Algorithm

CONTENTS

Plant breeding is the science and art of changing the traits of the plants to get the desired characteristics in the new plant. The first half of the chapter tells about the meaning, and objectives of plant breeding. The meaning of pleiotropy and its mechanisms are also explained in this part along with the meaning of ideotype.

The mid part of the chapter is about the methods of selection in the crop improvement. The part also discusses about the concept of domestication and the domestication syndrome. The last part of the chapter gives insights about the seed quality and the factors that affect the seed quality.

3.1. INTRODUCTION

There are several techniques through which breeding in plants can be done. These techniques may range from simply selecting plants with desirable characteristics for propagation and it may go beyond complex molecular techniques. It was since the beginning of the human civilization that plant breeding is being practiced and it is nearly thousands of yours. Now, plant breeding is practiced by individuals worldwide. There are gardeners and farmers, professional plant breeders who are the employees of the organizations such as crop-specific industry associations or research centers, government institutions, and universities.

The breeding of a new plant is important for ensuring food security by developing new varieties as per International development agencies. Plant breeding helps in higher-yields, drought-resistant, resistant to pests and diseases, and regionally adapted to different environments in different growing conditions.

The objectives of this may be as follows:

- Improved characters in agronomy
- Crop improvement
- Resistance against biotic and abiotic stress

3.2. OBJECTIVES OF PLANT BREEDING

3.2.1. Higher Yield

Improvement in the field of "Economic produce on economic part" is the main aim of plant breeding. Depending upon the crop species, it could be fodder yield, tuber yield, fiber yield, cane yield, oil yield or grain yield. This can be achieved by evolving the high yield variety or hybrids.

3.2.2. Improved Quality

One of the important objectives of plant breeding is the quality of the production. However, the characters of quality vary from crop to crop. For example, the quality in wheat could be its size, milling, its color and baking quality.

The quality in rice could be its cooking quality, invalid could be malting quality, vegetables its color size nutrition and keeping safe quality, the protein quality in pulses, content of oil in oil seeds, strength, fineness, fiber length in cotton.

3.2.3. Abiotic Resistance

Abiotic factors such as heat, wind, drought, soil salinity, cold, extreme temperatures and frost affect the crop plants. In order to protect the crops from such factors, breeders are required to develop the resistant varieties for such environmental conditions.

Figure 3.1: Drought adversely affecting the crop.

Source: Image by USDA

3.2.4. Biotic Resistance

There are various diseases and insects that affects the crop production, which leads to losses in the yield of the crops. For overcoming these challenges, the best method which is also the cheapest one is the genetic resistance method that helps in minimizing such losses. Breeders can develop resistant varieties through the use of resistant donor parents which is there in the gene pool.

Figure 3.2: Crops affected by insects.

Source: Image by Wikipedia

3.2.5. Change in Maturity Duration / Earliness

The most desirable character of the plant breeding is the earliness that has several advantages. This method is equipped with less insecticidal sprays, permits new crop rotations, less crop management period and it can extend the crop area. The rotation of the rice-wheat crop has been permitted by the late planting with the development of wheat varieties.

Therefore, the duration of the crop maturing is the important objective for the breeder. The crops that are maturing early, the crop varieties, or varieties that are suitable for different dates of planting has to be seen. In cotton, the duration of the maturity has been reduced from 270 days to 170 days, 360 days to 270 days in sugarcane and 270 days to 120 days in pigeon pea.

Crops such as pigeon pea (*Cajanus Cajan*), mung, cotton (*Gossypium sp.*), etc. requires development of varieties having determinate growth in the above crops.

3.2.6. Dormancy

If there are rains at the time of maturity, some crops may show results as the seeds germinate even before harvesting in the standing crop. For e.g., Barley, Green Gram, Black Gram and Pea, etc. For checking that how much los has occurred due to germination. There is a need to introduce a period of dormancy in these crops to check loss due to germination. However, there may arise a need to remove dormancy.

3.2.7. Desirable Agronomic Characteristics

Some of the desirable agronomic characteristics are tilling capacity, plant height, trailing habit, branching, growth habit and erect etc. For example, in crops like cereal, dwarfness is generally related to lodging resistance and response of better fertilizer. In fodder crops, tallness, profuse branching, and high tilling.

3.2.8. Elimination of Toxic Substances

The varieties of some of the food crops has to be grown so as to keep them away from the toxic compounds that are harmful for human consumption. Some of the examples of the harmful effects of the toxic substances are explained further.

Presence of neurotoxin in Khesari – lentil (*Lathyruys sativus*) can result in the paralysis of lower limbs, presence of erucic acid from *Brassica* which can harm human health, and removal of the gossypol from the seed of cotton which is necessary to make it fit for human consumption. Therefore, the nutritional value of the crops would increase with the removal of the toxic substances.

3.2.9. Non-Shattering Characteristics

In crops such as green gram, the shattering of pods is a serious issue. Therefore, it becomes necessary to shattering resistant in crop like green gram.

3.2.10. Synchronous Maturity

The maturity of the crop species at the same time is called as Synchronous Maturity. This characteristic is highly desirable in the crops like castor, cotton, cowpea, and green gram. The reason being, in these crops several pickings are done at the time of crop harvest.

3.2.11. Photo and Thermo Insensitivity

The cultivation boundaries of crop plants can be crossed with the development of the varieties that are insensitive to light and temperature. The Photo and thermo-insensitivity have made it possible for producing the wheat and rice varieties in new areas.

3.2.12. Wider Adaptability

Adaptability is the situation where considering the wide range of environmental conditions it is suitable to cultivate a variety of crops. Therefore, adaptability is considered as an important objective in plant breeding. This is because, this characteristic helps in the stabilization of the crop production over regions and seasons.

3.2.13. Varieties for New Seasons

Maize is considered as a kharif crop traditionally. Nowadays, scientists are growing maize as rabi and Zaid crops. Similarly, in addition to the main kharif crop, mung is grown as a summer crop.

3.3. PLEIOTROPY

Generally, pleiotropy refers to a situation where a single gene mutation causes phenotyping changes in multiple distinct traits. These changes are considered as large phenotypic changes which are often deleterious in individuals.

In pathological syndromes the pleiotropy effects were first recognized as observed in phenylketonuria. While if it is studied across individual population for manifest, as correlation between traits it can leave minute effects on multiple traits.

The genetic structures which is developing can be revealed by the pattern of mutational effect on the phenotype traits. As it suggests that the treats are shared if there are simultaneous changes in different traits of the portion of genetic development.

Ludwig Plate a developmental geneticist, was the first person who coined the term pleiotropy in the year 1910. Plate was interested in the intersection of evolution and development. Also, the effects of pleiotropy were recognized earlier in the correlated effects and medical syndrome by Mendel and Darwin.

Figure 3.3: Ludwig Plate, who coined the term Pleiotropy.

Source: Image by Wikipedia

There were many attempts made to explain the physiological basis of pleiotropy in many of the earlier studies (Stearns, 2010). Yet there is an argument that provides the strongest impulse for the making of the studies of pleiotropy in the evolutionary biology like that of culminated in morphological integration.

However, it is the work of the phenotype correlation and their consequences for the revolutionary change that has made it possible.

The quantitative genetic study of evolutionary consequences of the underlying genetic covariances and the genetic covariances has been motivated by study of phenotypic correlations and covariances (Lande, 1979), the relationship between the genetic and phenotypic correlations (Cheverud, 1988) and the effects of pleiotropy on standing variation (Turelli, 1985).

The genetic variation direction along with the particular directions of phenotype are the main evolutionary consequence of pleiotropy and it includes preventing others also. It is seen that there is a biasness toward those phenotypic directions in evolutionary change by those that exhibit the greatest heritable phenotypic variance (Schluter, 1996).

There can be either prevention or diminishing response to selection by the variation shared with other traits with regards to any particular trait. Also, it is found that there can be correlated response in other traits.

When there is more than one trait which are affected by the mutation at a locus, however, the only advantageous step is the change in a single trait. This is dependent on the selection regime acting on all affected traits, and choosing whether such variation can be selected or not.

The response in a population is likely to be not realized if such correlated effects are basically deleterious. Despite selecting the sufficient variation in the selected trait (Hansen et al. 2003). The above stated situation is known to be as evolutionary or pleiotropic constraint.

Other traits carrying the neutral attitude by the correlated mutational effects, there is a possibility that an advantageous mutation can spread in a population. Also, there will be change of traits when they are not selected directly. The model by George C. Williams (1957) is one of the most important examples using such pleiotropic side effects which states that antagonistic pleiotropy is the reason behind the senescence.

According to the model of senescence, the contribution to the individual's reproductive success which is contributed beneficial effects on individuals early in life are therefore selected and associated with the negative pleiotropic effect on late life stages which is referred to as senescence.

In the life period of the reproductive cycle there is a strong selection however it is not on the life period after reproduction (the cost of the reproductive fitness early in life outweighs by its gain). Although, in a recent time, there is a selective pressure which is proposed by the deleterious side effects and it has been compensated by secondary mutations.

In some of the conditions, one of the good example are the side effects of mutations that confers drug resistance, but in others, it works as reduction in fitness (Lenski 1988 and many later). The traits which are under the stabilizing selection, this model is suggested by the enrichment of compensatory mutations in traits. However, they share genes with traits under directional selection.

The traits' basis here allows the avoidance of the correlated response by the compensatory mutations. Basically, it means that instead of adopting a new phenotype, maintaining a new phenotype is advisable. This phenomenon underlies in the developmental systems drift (True and Haag 2001).

The developmental systems drift is a situation where the developmental processes are highly modified which are underlying the same trait in different taxa. While the end phenotype is under stabilizing selection, it appears that they are drifting (Johnson and Porter 2007; Pavlicev and Wagner 2012).

3.4. IDEOTYPE

The concept of Ideotype was coined by Donald, 1968. Ideotype concept is "a form denoting an idea". This was started as an aid to breeding programs related with the crops. This is with regards to the physiological factors that confers close in the first instance with the optimum yield in wheat (*Triticum Aestivum*).

There is an ideal model phenotype in effect which can perform in a given environment in a predictable way. This helps in the provision of the strategy and goal for genetic selection. Donald has derived the concept of crop ideotype from various other species.

Some of the examples are as follows, chickpea (*Cicer arietinum*) (Siddique and Sedgley, 1987), barley (*Hordeum vulgare*) (Rasmusson, 1987), fruit trees like mango (*Mangifera indica*) and apple (*Malus spp.*) (Dickman et al.1994), spruce and pine (Picea abies and Pinus sylvestris) (Karki and Tigerstedt, 1985) and forest trees (Dickman 1985).

The concept has evolved with the inclusion of trees in the concept. It has been described by the Dickman (1985) as 'the first step towards bioengineering an improved plant'. Under this, from the wild populations the selection of the potential breeding stock is done. Also, the selection is made by the combination of characters.

Therefore, it is important for the individuals in a wild population to survive. Especially, the long-lived perennials such as trees. Because they have evolved with an objective to survive and reproduce which is a different strategy from that of the plant breeders for enhancing the performance under the cultivation.

Therefore, it can be said that there is rare 'ideal' combination of traits in a crop plant which is derived from trees in the individual genotypes of the wild population. Dickman (1985) has emphasized on 'an ideotype is not a wistful construction born of unproven assumptions or opinions; it is rather a deductive product founded on a detailed understanding of plant morphology and physiology'.

Take for example, competition and sink dynamics; dry matter partitioning; phenology, metabolism of secondary compounds, and nutrient storage, etc. This ideotype breeding has been successfully been applied in many domesticated perennial species. Nowadays, this method is being applied in domestication of perennial tree and shrub species.

3.5. SELECTION IN CROP IMPROVEMENT

Since the very beginning of the human life, humans are working as a plant breeder. So, in order to increase the productivity of the crops, they have implemented a selection pressure on the crops for improving the productivity and assuring the stability in crops. The breeders of the plants have also done quantitative selection as for the purpose of the production of the crop species suitable for the particular uses.

A crop foe example maize, has been an evident that it is suited for boiling, popping, and cooking. The selection of the variants for ceremonial use has been done and retained as discrete types. One more instance for the influence of the human selection for the preference in crop domestication is choosing the particular crop over the other crop for main use.

Humans has selected wheat as the min cereal for consumption over the other crops such as barley, oats, cereal rye which serve as the similar characteristic as the wheat having the same adaptiveness. This is taken to be the other example for human discernment in selection.

The reason behind the selection of the wheat as the main crop in the palate preference is that it contains a protein complex, called gluten. This is because wheat contains the unique physical properties such as extensibility and resilience. This as a result produces a bread which is more pleasing to the palate than any other cereals.

Mutation is the newest source of all genetic variability in the wild. The mutant genes are first selected and if they are found to be advantageous, it is produced through hybridization into the cultivated forms of the species, which is a rare case in the self-pollinators, but is common in cross-pollinators.

Inbreeding crop contains a large numbers of genetically-fixed lines. While, as an out crosser, the genetically diverse populations is always present. There is maximization of both the genetic homogeneity and genetic heterozygosity in a special case of F_1-hybrid varieties.

In both self-pollinated and cross-pollinated plants, there are two types of methods for the selection. One of the methods is similar called as mass selection. The advantage of this method is its significance in crop plant improvement. Throughout the history of crop domestication and the recent plant breeding programs has all the advantages.

For the self-pollinated plants, the method of selection is known as pure line selection. The method of selection for cross pollinated plants is the progeny selection and line breeding. This method basically selects the single plants from the crop populations. Their progenies are evaluated for superior performance.

This selection can modify or form the desired lines or varieties that are genetically needed for the character(s) being selected. There is difference in implications and procedures in selection, owing to the different population structures in self-pollinated and cross-pollinated crops.

Because, in the self-pollinated crop plants, the homozygous condition is achieved through selfing. The production of the homozygous pure line varieties can be done from the single plant selection. In a cross-pollinated crop, there will be a consequent loss in fertility and vigor, and the selection of the single plant would lead to inbreeding and homozygosity.

This would be the result of the heterozygous nature of all the individual plants in the population. Therefore, the selection of the latter crops has to be based on the population of the plant where there is a high-level of out-breeding and hence heterozygosity maintained.

3.5.1. Mass Selection

According to this method of selection, the group of individuals are selected from the population based on the similarity in the phenotype. This method aims to improve the performance for the character or characters by the population.

The breeder of the crop is indicated by the extent of the population variability for the character and knowledge of the genetic basis. The crop breeder gets the indication of the potential offered for improvement through selection.

In contrast to the methods of the pure line selection, progeny selection and line breeding, the mass selection procedure has no progeny test. However, in order to gauge the effectiveness of selection, the evaluation of the selected population is done against the original unselected population.

The mass selection method in self-pollinated crops has two main uses. First, this method brings a rapid improvement to the population in an efficient manner. It selects a population of the individuals that are not fit absolutely i.e., visually. Simply, they may be less matured or too late matured, lodging or disease susceptible.

Therefore, the mass selection method is used to 'purify' the populations for a narrower range of character value(s). This means that the method basically for the commercial production, is believed to confer closer-to-optimum fitness for that character or characters.

In a self-pollinated crop variety, it is also used for the production of the pure seed. In this, there is selection based on the discarding phenotypic variants. It is diverged from the standard of the varieties or range. Also, the mass selection method has proven to be a best method for the cross-pollinated crops as it has contributed substantially in the improvement of the crop. The improvement is particularly for characters of high heritability, for example, date of maturity, disease resistance, and protein or oil content in grain.

This method is reported as the best method to be applied successfully. However, like every other method, this method is also carrying a disadvantage with regards to the improvement in the characters of the low heritability such as yield.

3.5.2. Pure Line Selection

In the self-pollinated plant species, there is a continued selfing of a heterozygous individual or population. This will result in the further population to become homozygous in an increasing proportion. This is the knowledge that the pure line selection method depends on.

A time will come when all the individuals will become homozygous. The parents of a crop in a crop breeding is differed by a large number of pairs of gene. This number will however influence the number of generations of reaching the homozygosity for which self-fertilization is necessary for the population which is hybrid.

The population becomes heterogeneous at this point of time which is comprised of a large number of different homozygous individuals. The number of generations of selfing since hybridization and the number of independent gene pairs through which there is difference in the parents. The formula for the proportion of the population which is homozygous is as follows:

$((2^m-1) / 2^m)^n$

Where, m = the number of generations of self-fertilization and,

n = the number of independent gene pairs by which the parents differ.

Therefore, it can be said that the proportion off the homework zygosity after 6 generations of the hybrid population of self-fertilization would come out to be approximately 83%, if there are 12 independent gene pairs of the parents of a cross. Inbreeding programs the most used method is the pure line method of selection in order to improve the self-pollinated crops such as Barley and wheat.

3.5.3. Progeny Selection and Line Breeding

Over the last 70 years, the progeny selection and line breeding method has come out to be an important method in improving the performance of the cross-pollinated crops. Under this method the plants are selected on the basis of its phenotypic appearance from the open pollinated populations.

After that the selfing of plants is done and then the evaluation of their progenies is carried out in usually small rose for plots of around 10 to 50 plants in the coming years.

There is elimination of the inferior progenies and incorporation of the selected progenies into a composite population. For the character of characters selected that composite population must be genetically superior to the original population.

The line breeding method is basically the procedure that incorporates the selected superior progenies into the composite population. However, the risk of inbreeding can be reduced. There are two ways through which it can be done, one way is to incorporate a large number of progenies usually which are greater than 50 into a composite population and the other way is to ensure of its dissimilarity in genetics.

The progeny and line breeding method has its own advantage as it has proved useful for the improvement of the performance of cross-pollinated crops for qualitative characters and for quantitative characters as well.

However, this method is not suitable if the population is highly adopted and hence it will not serve as an efficient method for improving yield the reasons being the low heritability and complex genetic makeup of the yield character.

3.5.4. Natural Selection

For the purpose of character improvement, the breeding programs in self-pollinated and cross-pollinated crops is dependent on the artificial selection. There has been some interest received for the improvement in crop breeding by the use of such natural selection.

Generally, the uses of the natural selection method apply mainly to adaptive characters such as disease resistance or tolerance and cold or drought tolerance. Basically, the Natural selection method acts on large, segregating populations of self-pollinated crops.

For example, the production of crops for a number of seasons in a particular environment of the 'composite cross' populations of barley, and normal populations of cross-pollinating crops. Under the natural selection method of cop selection, there is increase in the reliability of the yield of the crop in an environment. There is a significant shift obtained in gene frequency which gives greater adaptiveness.

3.6. THE CONCEPT OF DOMESTICATION AND THE DOMESTICATION SYNDROME

Domestication would be, fundamentally, a selection process of evolution for adapting to the transformed environmental conditions, since not just for consumption, plants would also be used for cultivation. For the occurrence of domestication, cultivation would be considered as an essential norm. To a great extent the forces of selection acting on plant population in environments that would be original and natural, would be varied from the following cultivated environments.

Among plants, a tough intra and interspecific competition would be existing in the natural environments for light, water, and nutrients of the soil. Whereas, there would be less competition in the agroecosystem, even in prominent diversified arrangements, such as agroforestry (Leakey, 1998; Perfecto and Vander Meer, 2010) as well as gardens (e.g., Ban and Coomes, 2004; Buchmann, 2009).

Such minimized competition would be transforming into increased productivity for harvested organs (e.g., grains, leaves, roots, etc.) by means of photosynthates partitioning being modified along with more substantial index for harvesting (ratio of harvested biomass to total biomass). The developmental requirement for a more competent production system, related to biomass per unit of surface and time, would be another prime difference between natural and cultivated environments.

The entire objective would be towards expanding the harvest biomass, even though lowering the total effort required for the biomass harvesting. Thus, for the accomplishment of the goal, many traits are supposed to be selected. Such traits would include:

- Seed dormancy loss leading to more even stands at planting
- Growth habit that would be additionally compact, with lower branching and shorter branches reducing intraspecific competition providing a more synchronous harvesting
- A tendency towards evolving reproduction from outcrossing to inbreeding to vegetative propagation so as to reduce the dependency of reproduction on abiotic and biotic factors
- Minimized or decline of seed shattering which would be limiting loss of grains during harvest.

Third, would be application of an important selective role by humans not solely as farmers but also as consumers of their own production, regardless of whether it would be for food, feed or fibre , in such case an important feature would be human attraction for variations, like new seed and colors or shapes of fruits. It would lead to outstanding morphological diversity in few crops, like developmental changes in cabbages (*Brassica spp.*: red cabbage, Brussels sprouts, cauliflower, etc.); some legumes with varied seed color, size, and shapes (*Phaseolus sp.* and *Vigna sp.*); some fruits with different size, shape, texture and color in the Rosacea (e.g., apple: *Malus sp.*; pear: *Pyrus sp.*; and cherry and plum: *Prunus sp.*).

Regardless of whether it would be roots, leaves, fruits, or grains, normally, domestication would be resulting in the size of the harvested organs being expanded. Expansion of size at time would be outstanding due to their achieving a difference of 10x-20x between wild and domesticated varieties. Usually plant would have toxic elements for accomplishing various functions like disease or pest protection.

The concentration of such compounds would be usually lowering in all or to some extent part of the domesticated gene pool, for example, as depicted by cyanogenic glucosides reduction in cassava (Bradbury et al., 2013) or lima bean (Vanderborght, 1979). The occurrence of such groups of associated traits being the characteristic of domesticated plants along with showing certain level of uniformity among various, as well as disparate, crops resulted in calling them "domestication syndromes" by Hammer (1984).

Usually such syndromes would be appropriate for more vigorously domesticated plants; yet, on certain occasions major variances among crop plants would occur on the basis of their life history and mode of reproduction (Harlan et al., 1973; Smartt, 1990; Ladizinsky, 1998; Hancock, 2004; McKey et al., 2010, 2012). Additional limited differences would be displayed between wild and domesticated varieties, in perennial plants due to their prolonged juvenile phase, extended life cycle, high levels of gene flow and effective towards vegetative reproduction (Miller and Gross, 2011).

For example, in grape vine, the key trait of domestication would be evolving from a diecious (separate male and female individual) reproduction to an autogamous reproduction facilitated by hermaphroditic flowers (with both male and female reproductive organs). A single locus would be regulating the change (Battilana et al., 2013). Hence, usually, perennial crops domestic syndrome would be comprising of less differences between wild progenitor and domesticated descendant (Pickersgill, 2007).

The essential traits expression would also be frequently limited. There might be various syndromes of domestication for a particular crop, based on the conditions of local production. For example, for growing maize for purpose of grains (as would be the prevalent situation), a different domestication syndrome could be forecast, or in the form of biofuels feedstock for its vegetative matter.

A more branched type of plant is supposed to be involved in the latter, increasing the stems and leaves production than grain. An elaborate overview in relation to the extensive range of traits characterizing domesticated plants are considered to have been provided by Meyer et al. (2012). A discussion on the domestication level would have been provided by Dempewolf et al. (2008).

3.6.1. A Common Feature among Domestications: Reduction of Genetic Diversity at the Genome Level

Reduction of genetic diversity within the domesticated gene pool in comparison to the wild progenitor gene pool, would be a common feature of domestication. The level of reduction would be dependent on the species, their reproductive system and the molecular marker system that would be used.

In comparison to selfing species, the outcrossing species would be showing a lesser reduction of gene pool (e.g., maize: 83% nucleotide diversity retained in domesticated gene pool, Hufford et al., 2012, 2013; rice: 60–65%, Caicedo et al., 2007, or 35%, Zhu et al., 2007; common bean:B40% each in Andean and Mesoamerican domestications, Bitocchi et al., 2013; soybean:B60%, Lam et al., 2010).

High levels of genetic diversity would be maintained by vegetatively propagated crops as demonstrated by grape and apple (95% retention of diversity, Myles et al., 2011; Cornille et al., 2013). Propagation techniques related to distinct phenotypic variants, like grafting or cutting, along with the role of main genes in the genetic control of individual traits and existence of juvenility period and the extended period of production in vine or tree crops, could possibly lower the genetic diversity.

However, genes would be flowing from local wild population, small-scale propagation, local adaptation to environmental conditions and preferences of consumers maintaining extreme levels of diversity (Cornille et al., 2013).

Multilocus association would also be affected by domestication, measuring as per linkage disequilibrium (LD), the association of gametes nonrandomly. The no randomness could be resulting from close association, wherein, LD could be used for mapping loci for regulating phenotypic traits.

However, there would be other factors for inducing LD, like geographical isolation, genetic bottlenecks, and lower recombination for inducing population structure.

For example, in regular bean, consisting of two geographically isolated taxa (Andean vs. Mesoamerican), the LD would be much higher in a sample that would have the two taxa (95% of locus pairs in LD) compared to in isolated samples of Andean and Mesoamerican (B70%; Kwak and Gepts, 2009).

Generally, LD would be increasing as a result of domestication (e.g., maize: Wright et al., 2005; Hufford et al., 2012; rice: Zhu et al., 2007; soybean: Hyten et al., 2007; sunflower: Kolkman et al., 2007; common bean: Kwak and Gepts, 2009; Rossi et al., 2009; foxtail millet: Wang et al., 2010).

3.6.2. The Genetic and Molecular Basis of the Domestication Syndrome

Crops and their wild progenitors would usually exhibit striking morphological and physiological variations due to adaptation-based selection to human cultivation along with food consumption purposes and non-food uses.

However, their biological species and their disparities are considered to have been originated few 10, 000 years before. The genetic origin resulting in such extreme transformation giving rise to crop plants as would be presently known is considered as being established recently. Originally, based on individual traits, genetic analysis would have been executed.

Usually, it would have two fold conclusion, specifically like, the majority of the domesticated traits would have been regulated through recessive alleles at one or two genes, signifying depletion of function mutations (Ladizinsky, 1985). Normally such studies would be limited, however, due to insufficiency of information on the gene's location inside plant genomes and their potential association.

The capability of developing associating maps along molecular markers of adequate density and distribution in crop genome, accompanied by the separating populations development from the crossing of methodically selected wild and domesticated genotypes, improved the situation providing a more comprehensive image (including linkage information) of the domestication syndrome inheritance.

A sequence of general conclusions related to the genetic control of the domestication syndromes architecture resulted over a wide range of crops, primarily annual, herbaceous plants (Gepts, 2004).

In conjunction with genes or less effective QTLs, individual traits would often be controlled by the major Mendelian genes or quantitative trait loci

(QTLs), as per the initial conclusions. The single trait analyses would be confirmed with such observation.

Second, occurrence of the phenotypic variation would be resulting majorly due to genetic instead of environmental effects that would be signifying high heritability. A single genetic control would be signified by the first two genetic control, facilitating and accelerating selection at the initial stages of domestication.

Third, a subset of genes regulating the domestication syndrome in several crops would be at least loosely associated, on a restricted number of chromosomes. There might have been a benefit from the latter situation, wherein, occurrence of hybridization would be either between wild and domesticated varieties or among emerging domesticates in the early phases of domestication with varied elemental lineages associated to domestication syndrome are supposed to be recombined.

Recovering the hybridization syndrome parts would be simpler through association, even though due to repulsive association the domestication syndrome assembly would not be prevented.

3.7. GERMPLASM

Germplasm would be regarded as a living tissue facilitating the growth of new plants. There could a possibility for it to be a seed or another part of plant – a leaf, a piece of stem, pollen or barely a few cells that would be capable of becoming a complete plant. A species' genetic makeup information would be contained within the germplasm, which would be a useful natural resource of plant diversity.

Uniformity would be beneficial for crop plants within a diversity, ensuring steady yield providing easier management. Nevertheless, genetic uniformity in crops specifically, would be vulnerable towards new pests and stresses.

Plant breeders would be bestowed with perpetual ability for developing new high yielding, high quality varieties capable of resisting pest, diseases and environmental stresses evolving regularly through genetic diversity of germplasm.

The main element of genetic diversity would be the sexual compatibility of wild species and landraces as well as the crop species of ancestral varieties, however, the shrinkage of land for the growth of wild and several other plant species and types would be the cause for their disappearance.

Hence, the community for plant science is considered to have been developing conservational programs for gathering, preserving, evaluating, cataloguing, and distributing germplasm for people across the world for their usage.

Farming could be regarded as the original biological technology (biotechnology) wherein, cultivation and harvesting of particular plants for food production would have been initiated by humans, some 10,000 years before, for producing food.

Plant breeders would have augmented the selection process, consistently increasing the crop yield, and improving quality, through growing knowledge of genetics.

Few of the new technologies would be comprising of:

- Genomics, the genome study of organisms
- Genotyping, the method to determine the genotype
- Induced mutation, structural change of DNA or chromosomes generated externally, leading to a noticeable or detectable trait change
- Phenomics, a field of study with respect to phenotypes characterization
- Proteomics, the extensive study of proteins, specifically their structures and functions.

3.8. WHAT IS SEED QUALITY

The primary input in the production of a crop is the quality of the seed. Seed is known to be that part of the development of the agriculture that helps the farmers in the agricultural research by the impacts of the investments.

The seed must reach the farmers in a state that yields the farmers a good production of the crops. As, the seed is the limiting factor potential of the crop production. This is the reason that the seed must be with the farmers in a good quality stage.

There are many factors the seed depends upon. The quality of the seed is a complex thing (Fougereux, 2000) There are four key attributes that are explicitly identified (Bishaw et al., 2007).

- Genetic Quality

This refers to the genetic potential of the variety that is inherent for higher yield, better quality of grain and greater tolerance to biotic or abiotic stresses.

- Physiological Quality

The seeding emergence of the potential germination and vigor establishment of the crops in the field.

- Physical Quality

This quality ensures the contamination free with other crop seeds, noxious, common, or parasitic weed seeds, discoloration, mechanical damage, and having the uniform seed size and seed weight.

- Health Quality

This ensures the absence of infection/infestation with seed-borne pests (bacteria, fungi, nematodes, viruses, insects, etc.).

3.8.1. Genetic Purity

It was reported by McNeil et al. (2007), in Western countries, trade and commerce in countries language, there can be identification of two types of lentils which is based on the cotyledon and seed coat color. Such as red lentil is made up of orange cotyledon and pale grey to dark seed coat and green lentil is made up of yellow cotyledon with green to brown seed coat.

However, in the various regions of Africa and Asia, who are the major producer of lentils, this terminology is not followed. Moreover, the wide range of seed coat colors and genotypes with pure green cotyledons can be made available for trading and consumption purposes (Emami and Sharma, 1996a, b; Sharma and Emami, 2002; Sharma).

After that, the Red and green lentils of existing grouping will become fully redundant. The quality of grain is totally dependent on its genetic and varietal purity. Also, the uniformity of grain which is traded in bulk has to be noted as there are differences in the preferences of the consumers.

Moreover, the quality of the grain gets deducted when the different cotyledon colors are certainly mixed with the lentil green. Using door shield having high genetic purity for sowing is the best procedure in maintaining the varietal purity practically. This is due to the reason that the crop stand in lentil is dense and the cotyledons are concealed with short plant canopy.

3.8.2. Physical Purity

In order to produce good quality of lentil using physically pure seed is always desirable. The major vehicle for spreading parasitic weed such as

Orobanche and Cuscuta and introducing the contaminated seed into new areas and fields (Rubiales et al.).

There is greater economic significance in the Middle East with respect to these two weeds. However, in the countries of the Orient, they are not a major issue. This is the reason as to why *Orobanche* and *Cuscuta*, are considered as quarantine pests. This is however the assumed reason for rejecting the lots of seed in most certification schemes.

The Cuscuta can be eliminated by the usage of dodder mill and magnetic separator machines. However, there is one disadvantage that it is not applicable for large-scale area in lentils and is less practical.

3.8.3. Germination

The seed of the lentil which is physically sound like most of the field crops germinates well under laboratory conditions. Also, for the prediction of the field emergence, the most reliable crop is the lentil (Makkawi et al., 1999). Therefore, it can be assumed that if there is a proper care given to the production of the lentil crop, then it would not be a major constraint in lentil production.

3.8.4. Seed Vigor

Seed vigor can be defined as the capacity of the seed in various field conditions that the lot of seed can germinate and produce normal seedlings. There are various factors that controls the seed vigor. It is basically a quantitative character that controls factors such as mechanical injury to embryo or seed coat, stage of maturity at the time of harvest, senescence, environment and nutrition of the mother plant and seed size and attack by pathogens.

3.8.5. Seed Health

Seed health is a serious issue in the establishment of the crop. It has an adverse effect on the process of germination and the seedling establishment. Dose adverse effects leads to low germination, severe loss of yield, crop failure in total and reduced crop establishment.

One of the seed borne is lentil Ascochyta blight, along with anthracnose and Botrytis can you cause the death of the seed and ultimately result in the seedling blight and seedling with necrosis. The plant stand gets reduced and the seedling which was infected die just after the emergence having significant sequence on the yield of the crop.

For example, *Ascochyta lentis* causes low seedling emergence in lentil (Morrall and Sheppard 1981). The plants of lentil from Ascochyta which is infected gives very few branches having small roots and shoots. It has also reduced vigor and seed yields is also very low (Kaiser and Hannan, 1987).

3.9. FACTORS AFFECTING SEED QUALITY

There are several factors that extensively affect the quality of the lentil seed. Such factors can be production environment handling operations during the production and agronomic practices. Receipt quality is highly affected by factors other than the genetic factors.

These includes during the crop growth the environmental conditions affects the seed quality, development of seed and maturity (nutrient deficiency water stress soil condition biotic stress extreme temperatures), mechanical damage during treatment cleaning harvesting conveying and packaging, and last but not the least improper storage of seeds that can be affected by insects temperature content of moisture. Genetic factors

3.9.1. Genetic Factors

Genetic factors are those factors that affects the seed quality differences exists among species and varieties. As per Lakhani et al. (1986), there is a genetic variability in seed vigor. According to Lakhani, there is presence of hard seed among genotypes of lentil.

It is also evident from the fact that the thickness of the seed coat has an adverse effect on the germination process and leads to emergence in lentil (Sarker et al., 1995).

As per the findings, there were differences found due to the treatment of the aging and as a response to it among lentil varieties (Makkawi and van Gastel, 2006). The results of the treatment indicated that the high-vigor varieties can withstanding the aging treatments for longer duration. In contrast to this, low-vigor varieties cannot withstand the aging treatment.

Thus, in order to select high vigor genotypes which can be stored with full effect in the breeding programs, the test that can be helpful is the vigor test. The quality of the seeds can also be checked with the physical integrity of that seed. Therefore, one of the principal components of legume seed quality that is inspired by harvesting is the Testa quality. The conditions of testa can be different as it is more influenced by mechanical damage.

The content of the moisture in seed determines the degree of cracking from a given handling treatment. If the seed is exposed to mechanical injury it gets insect infestation. It also leads to increase of water during storage. This ultimately affects the poor germination, poor seedling establishment and reduced vigor.

During the times of high temperatures and high humidity, the lentil seeds shows sensitivity to injury as per the physiological studies during imbibition (Makkawi et al., 2008).

3.9.2. Production Environments

During the period of pre-harvest, the conditions of climate greatly influence the quality of seed. In the semi-arid or drier areas, the production of the seed of lentils is of good quality. There is excessive growth of vegetation during the season of high humidity and high rainfall and it also result in the yield reduction and bad seed quality.

There can be reduction in the yield of crops and flower drop due to excessive drought or high temperatures at the time of flowering and the pod-fill period. The exposure of lentil to these conditions during the crop growth causes drought stresses as it is sensitive to heat.

It also leads to a considerable reduction in yield (Johansen et al., 1994). There can be a dramatic fall in the seed quality in high temperatures during the development of the seed in the field (Hampton and Coolbear, 1990).

3.9.3. Harvesting

Harvesting may influence the quality of seed by the methods and the time of harvesting. Any delay in harvesting may weakens the seeds' ability to grow. Hence it may grow worse and cause greater loss of quality since it also tends to increase the biological age of seed.

Also, delaying harvesting may cause a reduction in seed germination significantly as it tends to leave the mature seeds under weathering conditions. This was reported that the delay of 1 week in harvesting causes 20% reduction in seed germination (Ellis et al., 1987).

3.9.4. Mechanical Damage

Another reason for the impure quality of seed or decreasing quality of seed is Mechanical damage. It can be caused at any time whether harvesting, cleaning, or handling. Seeds that have convex surfaces, for example Lentil seeds are more vulnerable to mechanical damage than the seeds having round shape such as bean, peas, or chickpea (Muehlbauer et al., 1985).

Mechanical damage could affect the quality of seed to a level that may ultimately results in abnormal seedlings. It may also result in broken or cracked embryos, split or chipped seeds. The relation between mechanical damage and moisture content has also been established and studied .

It is said that the breakage percentage increases with a decrease in moisture content below 14% and when the moisture content is higher than 20% (Fougereux, 2000). This directly states the effect of moisture content on Mechanical damage, according to this, damage is more when the moisture content at the time of mechanical damage is low.

3.10. CONCLUSION

Plant Breeding basically ensures that the characteristics of plant are improved. The basic reason of the improved crop is that they become more desirable agronomically and economically. Therefore, the main objective of plant breeding is to make the crop commercially successful by developing such improved varieties of crop plants.

Generally, "a successful variety is one with total balance of traits and makes it more profitable for growers than any other one they might choose. This is why breeders are wary about emphasizing one trait to the exclusion of others" (Lewis and Christiansen,1982).

At the end, it can be assumed that there are certain crops that needs improvement in some specific traits of that may become a prior objective for various economic and agronomic reasons. Therefore, it is the crop that decides the specific objectives that greatly depends on the crop under consideration.

REFERENCES

1. Bangarwa, S., 2020. *Crop Ideotype - It's Features and Steps in Ideotype Breeding*. [online] Biotecharticles.com. Available at: <https://www.biotecharticles.com/Agriculture-Article/Crop-Ideotype-It-s-features-and-Steps-in-Ideotype-Breeding-4170.html> [Accessed 13 June 2020].
2. Bishaw, Z., Niane, A. and Gan, Y., n.d. Quality Seed Production. *Lentil*, [online] pp.349-383. Available at: <https://www.researchgate.net/publication/226739279_Quality_Seed_Production> [Accessed 13 June 2020].
3. Gepts, P., 2014. Domestication of Plants. *Encyclopedia of Agriculture and Food Systems*, [online] pp.474-486. Available at: <https://www.researchgate.net/publication/288208269_Domestication_of_Plants> [Accessed 13 June 2020].
4. LEAKEY, R. and PAGE, T., 2006. THE 'IDEOTYPE CONCEPT' AND ITS APPLICATION TO THE SELECTION OF CULTIVARS OF TREES PROVIDING AGROFORESTRY TREE PRODUCTS. *Forests, Trees and Livelihoods*, 16(1), pp.5-16.
5. n.d. *01. Aims and Objectives of Plant Breeding*. [eBook] Available at: <http://eagri.org/eagri50/GBPR211/lec01.pdf> [Accessed 13 June 2020].
6. n.d. *Chapter 4- Plant Breeding*. [eBook] Available at: <https://cdn.csu.edu.au/__data/assets/pdf_file/0007/2805559/Chapter4_LuckettHalloran.pdf> [Accessed 13 June 2020].
7. n.d. *Definition, Aim, Objectives and Scope of Plant Breeding*. [eBook] Available at: <http://wizardsolution.yolasite.com/resources/PBG-4311.pdf> [Accessed 13 June 2020].
8. Pavličev, M., 2016. Pleiotropy and Its Evolution: Connecting Evo-Devo and Population Genetics. *Evolutionary Developmental Biology*, [online] pp.1-10. Available at: <https://www.researchgate.net/publication/309128242_Pleiotropy_and_Its_Evolution_Connecting_Evo-Devo_and_Population_Genetics> [Accessed 13 June 2020].
9. Bishaw et al., 2007
10. Cheverud, 1988
11. Dickman et al., 1994
12. Dickman, 1985

13. Ellis et al., 1987
14. Emami and Sharma, 1996a
15. Fougereux, 2000
16. Fougereux, 2000
17. George C. Williams, 1957
18. Hampton and Coolbear, 1990
19. Hansen et al. 2003
20. Johansen et al., 1994
21. Johnson and Porter 2007
22. Kaiser and Hannan, 1987
23. Karki and Tigerstedt, 1985
24. Lakhani et al.1986
25. Lande, 1979
26. Lenski, 1988
27. Makkawi et al., 2008
28. Morrall and Sheppard 1981
29. Muehlbauer et al., 1985
30. Pavlicev and Wagner 2012
31. Rasmusson, 1987
32. Sarker et al., 1995
33. Schluter, 1996
34. Sharma and Emami, 2002
35. Siddique and Sedgley, 1987
36. Stearns, 2010
37. True and Haag 2001
38. Turelli, 1985

4

Crop Breeding Methods

CONTENTS

Contemporary farming could only yield an insubstantial portion of food for humans and animals. Improved management could result in increasing the yield, however, to a minimal extent. With the advent of molecular technologies in the recent times, the methods used by plant breeders for identifying and selecting their germplasm have supposedly been changed as currently, the genetic differences would be evaluated at the DNA level.

Comprehensive guidelines and tutorials would be offered through different methods and protocols, suitable for various requirements and capacity from small laboratories examining molecular markers on the basis of one-by-one to the progressively popularized high throughout protocols intended for high capacity laboratories.

The chapter would comprise of methodologies facilitating the genetic diversity generation and its characterization. Recently, there had been a rapid development in the field of biotechnology as a practical way of expediting success in plant breeding and developing the crops that would be economically vital. The most significant methods for achieving such objectives are described elaborately in the chapter.

4.1. INTRODUCTION

Breeding of plants could be regarded as a process that would coexist among humans and edible plants. Plants used for agricultural purpose, would have undergone changes brought about by people, in turn the human population emerged from the changes that were supported by the new plant types.

Abundant harvests from plants would serve in unleashing some time from people for the development purpose of art, handicraft, and science, ultimately resulting in the contemporary human life since ages.

Existence of civilization would not have been possible in the absence of agriculture. However, the civilized world would not be able to sustain without implementing / introduction of modern varieties of crop in agriculture (Harlan et al., 1992). Therefore, it would establish the fact that crop breeding would be regarded as the vital pillars of civilization.

Figure: 4.1: Crop Breeding.

Source: Image by Pikist

A handful of the population in industrialized nations would be involved in agricultural activities. Most of the people for survival purpose, would be dependent on a social arrangement strategy. According to the strategy, food would be provided by a certain group of people in return of few services or goods.

In modern life the strategy seems to be quite basic with the availability of food in the nearest supermarket and most of the time people would take the privilege for granted. Nevertheless, a catastrophe in the world of agriculture would lead to impediment of the above mentioned strategy resulting in scarcity of food among people. Hence, by applying protective measures in agriculture it would be possible to safeguard the strategic pillars of modern civilization.

The essence of breeding crops would be choosing the best categories from a group of variants. It would be pertinent to the producing and quality of edible portions; easy cultivation, harvesting and processing; withstanding environmental pressures; pests resilient.

Such individual aspects associated to agronomic or food value could be divided into several distinct traits. Each such trait would be delivering its own spectrum of variation. Engineering an individual trait and overlooking the rest others would be rather natural; however, getting a productive variety out of it would be quite doubtful.

Plant breeding challenge would be dwelling in improvising all traits of interest together with genetic correlations among various traits, making the task hard. It might be resulting from the genes with pleiotropic effect, to physical association among chromosomal genes, or in relation to population genetic structure (Hartl et al., 1997).

Adopting one trait would lead to the transformation of associated traits, occasionally in the intended direction and at other times in an adverse manner (Falconer et al., 1996). Hence, selection could result in unexpected transformations, usually within the variety that would be normally detected in crops, thus, with no expected threat for consumers or the environment. However, whether the theory provided, would be valid or not is considered to be a debatable topic (Kok et al., 2008).

In consideration to the rising standards of knowledge related to the fundamental mechanisms and regulating the production and selection process of better quality plant varieties, the main purpose would be to discuss the techniques of plant breeding in the path of technological evolution.

Three major eras of crop breeding could be identified in the particular context:

- Plant breeding on the basis of adopting observed variants, irrespective of their origin
- Production and selection of wide variation through regulated mating
- Inspection of within-genome inheritance variation and choosing the particular recombinants

4.2. EVALUATION OF PLANTS

The breeder is supposed to decide which specific plants would be abandoned and which to be approved for producing the progeny. Hence, estimating the plant value is considered to be a challenging task related to few traits, compared to others.

4.2.1. Qualitative Characters

The involvement of intermittent or qualitative differences administered by one or certain vital genes would be considered as the simplest characters, or traits required to be addressed. Several such innate disparities would prevail which would be deeply impacting the value of plant and its utilization.

Figure: 4.2: Green Beans.

Source: Image by Public Domain Pictures

For example, comparison between starchy and sugary kernels (property of field and sweet corn, respectively) and in green beans comparing the growth habit between determinant and indeterminant (determinant varieties would be accustomed to mechanical harvesting). Such disparities are supposed to be easily noticed and quickly assessed along with the expression of traits remain consistent irrespective of the plant growing environment. Such traits would be known as extremely heritable.

Figure: 4.3: Sweet Corn.

Source: Image by Flickr

4.2.2. Quantitative Characters

Nevertheless, in other cases, plant traits would be grading progressively from one radical to another in an unremitting series and categorizing them into separate groups would be impossible. Hence, the variability would be called quantitative. Several traits associated to economic benefit would be of such types. For example, height, cold and draught tolerance, maturity time and specifically, yield. Various genes would be governing the traits, with minor effect on each. Though there would be no absolute difference between the two types of traits, it would be however, suitable for designating qualitative characters like the ones related to specific differences and quantitative characters like those related to a graded series.

It would be more challenging for the breeder to regulate quantitative characters. There would be three essential reasons behind it:

- Hereditary change would be slowed down and difficult to measure by the total number of genes involved
- By measuring and exacting statistical analyses usually help to detect the trait variations involved
- Majority of the variations would be resulting from the environment instead of genetic endowment; for example, the heritability in few

traits would be lower than 5%, which means that genes would cause the 5% of the variation observed and for the rest 95% it would be the environmental causes.

As a result, thoroughly designed experiments would be essential to differentiate the superior plants. It is due to the fact that they would be carrying quality genes from the ones superior as they are supposed to be growing in a congenial site.

4.3. VARIOUS METHODS OF CROP BREEDING

4.3.1. Mating Systems

The type of pollination, or pollen transferable from one flower to another would be devolved through angiosperm mating. When pollen is transferred from one flower to another flower of the same plant, it would be termed as self-pollinated (a "selfer"). Whereas, if the pollen from a flower of a different plant is transferred, it would be called cross-pollinated (an "outcrosser" or "outbreeder").

Nearly 50% of the top cultivated plants are considered to be cross-pollinated naturally including several devices in their reproductive systems for cross-pollination. For example, protandry (shedding of pollen prior to ovules maturing, like in carrot and walnut), dioecy (male and female parts would be borne on different plants, like hops, date, asparagus and palm) along with genetically established self-incompatibility (pollens unable to grow on the same plant stigma, like white clover, cabbage and several other species).

Species of other plants comprising of a greater proportion of the top cultivated plants, like wheat, barley, rice, peas, beans, and tomatoes are mostly self-pollinating. Comparatively, the reproductive mechanisms would be less for promoting self-pollination; the most definitive being the flowers failing to open (cleistogamy), like in few violets.

The pollen would be shed prior to or right at the time when the flowers in case of barley, wheat, and lettuce. The opening of the flower would be followed by pollination in tomatoes; however, surrounding the stigma a cone would be formed by the stamens. Unwanted cross-pollination risk would always prevail in such species.

Pollen of the chosen male parent, excluding any other pollen, reaching the stigma of the female parent would be mandatory in controlled breeding procedures. With stamens and pistils existing in the same flower, the selected female flower anthers are supposed to be removed prior to the shedding of the pollen. It would generally be done through forceps scissors.

It would also be crucial to protect against "foreign pollen". Using a plastic or paper bag for covering the flower would be the most general technique. Transferring of the pollen from the chosen male parent would be done on the female parent stigma becoming receptive. It would be done by often breaking an anther on the stigma and replacing the protective bag.

Hence, few hybrids production would be laborious and costly as it would frequently need a sequence of fine, exacting as well as well-timed hand operations. Controlled breeding would be simpler in case of male and female flowers occurring in different flowers, like in corn (maize).

With two parents in a cross-pollinated plant, each probably varying in several genes, would be producing several traits of diversified hybrid (heterozygous) plant populations. A single parent self-pollinated plant would be producing a more stable pure breeding (homozygous) plant population for various traits.

Therefore, self-breeders would probably be considered to be extremely homozygous in comparison to outbreeders. Hence, it would be the right breeding for a particular trait.

4.3.2. Self-Pollinated Crop Breeding

With self-pollinated species, the successful breeding methods would be:

- Mass selection
- Pure-line selection
- Hybridization, the pedigree method, the bulk method, or the backcross method handling the segregation generations
- Development of hybrid varieties

Mass Selection

Seeds would be collected from (generally from a few dozen to a few hundred) individuals that would appear desirable, among a population, sowing the following generation from a plethora of assorted seeds. The mechanism often denoted as phenotypic selection, would be related to the individual appearance.

For improving old "land" varieties, mass selection is supposed to have been extensively used. They would be the varieties that are considered to have been transferred from one generation of farmers to the following, since ages.

Destroying the undesirable types in the field for the purpose of simply eliminating them, would be an alternative approach which is supposed to have undoubtedly been applied for so many years. Irrespective of the fact that whether it would be the saving of superior plants or elimination of the inferior plants the results would be alike: the following season would have a stock of the superior plant seeds for the purpose of plantation.

Separately harvesting the best plants as well as growing and comparing their progenies would be a contemporary purification related to mass selection. The inferior progenies would be destroyed harvesting the seeds belonging to the remainder.

It would be worth noting that presently mere appearance of the parent plant would not determine the selection alone, however, the look/phenotype and performance of the progeny would also be a major concern for selection.

While dealing with quantitative characters of low heritability, progeny selection would be considered more potent, compared to phenotypic selection. It supposed to be kept in mind that since an extra generation would be required for testing the progeny; therefore, there would be double gain for every individual selection cycle in comparison to simple phenotypic selection for gaining an equivalent rate of gain for every single unit time.

Probably the easiest and inexpensive type of plant-breeding procedures would be mass selection with or without testing of the progeny. It would be extensively used for breeding few forage spices, not that economically essential for supporting more comprehensive attention.

Pure-line Selection

Pure-line selection is usually accompanied by three more or less specific steps:

- Genetically variable population would provide several superior looking plants for selection
- Growing the progenies concerning separate plant selections and by simple observation throughout a period of many years, evaluating them

- Extensive trials that would involve cautious measurements for ascertaining the left over selections in terms of best yielding abilities and other performance aspects, would be incorporated selection would not be possible as per observation solely.

Following, any progeny superior in comparison to the prevailing variety, would be released in the form of a new "pure-line" variety. During the initial phase of 1900s, the dependency of the major portion of success related to the method would be on the availability of genetically variable varieties of land waiting to be exploited. They would be delivering a supply of superior pure-line varieties, few of them however, being represented within the commercial varieties.

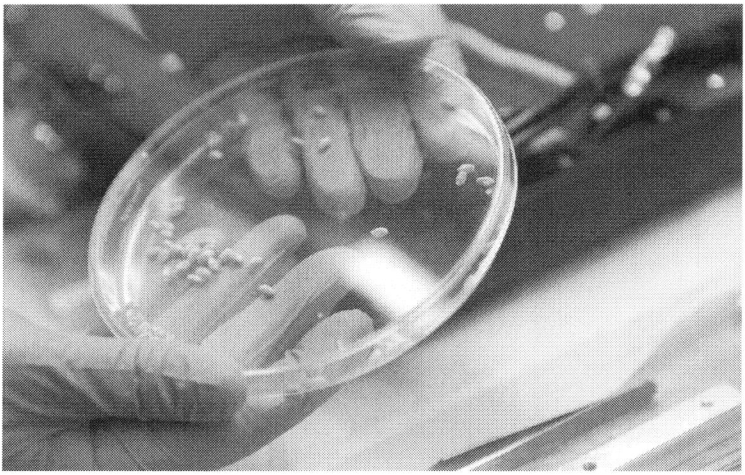

Figure: 4.4: Selected Seeds Used for Growing Crops.

Source: Image by Picyrl

The above mentioned pure-line method is supposed to have been experiencing a diminishing value for breeding vital cultivated species in the recent times. Nevertheless, the inferior species that would not be largely selected, are considered to be using this method extensively.

The selection of single-chance variants, mutations or "sports" within the original variety would be the centuries old variation of pure-line selection method. Characterization of extensive varieties differing from the original strain, like color, absence of thorns or barbs, dwarfness and resistance towards disease, would result originating in such manner.

Hybridization

Breeding self-pollinated species by the method of planned hybridization between accurately chosen parents is considered to have become paramount in the 20th century. The hybridization goal would be to merge desirable genes present in two or greater varieties, producing superior progeny of pure-breeding in various regards to the parental types.

Genes, after all, are supposed to be accompanying other genes invariably, in a group known as a genotype. One of the biggest obstacles faced by the plant breeder would be handling the huge number of genotypes in an efficient manner, occurring in the following generations of the hybridization. For example, a cross pollination between hypothetical varieties of wheat with a difference of mere 21 genes would be able to produce greater than 10,000,000,000 various genotypes in the second generation.

For growing a population as big as to allow every genotype occurrence in its probable frequency, farmers would require normal spacing at more than 50,000,000 acres. Statistically it would be possible for 2,097,152 various pure-breeding (homozygous) genotypes occurring, each being likely a new pure-line variety.

In contrast to the huge majority of the second generation genotypes this would be considered as hybrids (heterozygous) for one or several qualities. The importance of competent approaches to manage hybrid population would be illustrated by the numbers, due to which the pedigree procedure would be most extensively used.

Pedigree Breeding

The two genotypes crossing, each with a single or many desirable characters that might be lacking in the others pedigree breeding is supposed to begin. The inclusion of a third character could be possible if all the desired characters would not be provided by the two original parents, through the adaption of one of the hybrid progeny related to the first generation (F_1) by means of crossing. Selection of superior types would be done in the following generations, in the pedigree method, along with the maintenance of parent-progeny relationships record.

The huge majority associated with the released cultivars in relation to self-pollinated species are considered to have been established by means of pedigree method. Pedigree breeding would be embodying the crossing of parents and creating separate populations, being executed through several generations of self-pollination and selection till the time a combination of good characteristics of both parents as well as a set of derived lines would be obtained.

The said method would be effective for qualitative traits breeding since it is supposed to be formulated on the complementation of traits. For example, resistances of disease, or traits being easily classified, like architecture of a plant or the color or shape of different parts of a plant.

The pedigree method providing for building superior varieties by organizing, within the same plant, good characteristics found in different materials, would be appealing for breeders. All crossings being regulated, knowing the genealogy of every individual cultivar would be possible.

Efficient evaluation of the yield solely at the process end, on inbred lines, at the time the seeds being available for replicative/explicative trials, is considered to be the major weakness related to the pedigree method. At such a point, there would be negligible room for improving the potential of the yield though until there would be an advancement of a large number of lines.

As a result, the progress rate for resultant yield from the pedigree method would be usually moderate, occasionally beyond 1% per year (Breseghello et al., 2011).

Bulk Population Method

The handling of generations subsequent to hybridization would be the major factor that would differentiate bulk-population breeding method from the pedigree method. At regular commercial planting rates, the F_2 generation would be sown, in huge plot.

A mass harvest would be conducted once the crop is matured, using the seeds for establishing the subsequent generation in a huge plot. No ancestral records would be maintained.

Plants with low survival value are considered to be eliminated at the time of bulk propagation natural selection. Two artificial selection types usually enforced would be:

- Destroying plants carrying unfavorable primary genes
- Mass procedures, like harvesting wherein, a part of the seeds would be mature, selecting early matured plants or using screens for selecting increased size of seeds.

In a similar way to the pedigree breeding method, selection of single plants would hence, be conducted and evaluated. The breeder handling numerous individuals economically, would be the major advantage of the bulk-population method.

Figure: 4.5: The Best Variety of Seeds Would be Chosen for Bulk Population Method.

Source: Image by Public Domain Pictures

Improvisation of an exceptional variety can be done through the transfer of few distinct desirable characters in it, which it might be deficient in. Hence, by crossing a superior variety of plant first, with a donor variety plant, carrying the related traits followed by the mating of the progeny back to a plant with superior parent genotype, it could be achieved. The method would be known as backcrossing.

The progeny would be hybrid, following five or six backcrosses, for transferring of the character, however, similar to the superior parent related to all other genes. Selfing final backcross generation, together with selection, is supposed to be producing few progenies pure breeding for the transferred genes.

The benefits of backcross method would include its rapidity, requirement of a smaller number of plants and the outcome predictability. The occurrence of the chance coupling of genes, leading to outstanding performance improvements, would be diminished by the procedure, and considered as a major drawback.

Hybrid Varieties

In comparison to inbred individuals, heterosis would be considered as superior individual (Shull, 1948). Parents being the more divergent, within a few ranges, with heterosis higher on their offspring (Moll et al., 1965, Reif et al., 2005).

Inbreeding for generations leading to rapid decay of hybrid vigor as an indication of whatsoever might be the underlying heterosis mechanism would be due to the heterozygous loci being present. Therefore, the main focus of the maize breeding programs is presently on the development of competitive F_1 hybrids, wherein, heterozygosis would be at its highest level.

Two challenges would accompany the hybrid programs:

- Improvisation requirement for a minimum of two populations concerning agronomic adaptation, simultaneously placing them genetically far enough for the expression of strong heterosis
- Development of coherent seed production related to selected hybrids, in an attempt of not offsetting the additional yield value by the seed production expenses, from the resultant heterosis.

The first concept of heterotic groups would be considered as a resultant of the first challenge in case of maize (Melchinger, et al., 1998). The elite gene pool would be split into subsets, applying population breeding within it.

By improving the parental lines leading to more yields in field of seed production, using male sterility genes and mechanical emasculation ("detasselling") for the purpose of bypassing self-pollination, the second problem would be solved.

Alternatively, from the business point of view, greater benefits would be presented by hybrids. The hybrid plants producing seeds would be genetically heterogeneous through thousands of genes being segregated.

A huge difference in agronomic traits, architecture of plant and duration of the cycle would be presented by the resulting crop produced by the seeds from hybrid plants, thus, lowering the yield and quality of grain.

Hence, farmers would be advised to purchase new seeds every year that would lead to a relentless demand for hybrid seeds. Further, with transgenics emerging, the carriers of the important proprietary traits would preferable be the hybrid seeds. It would be due to the fact that controlling of the event by the owner would be better supported by the hybrid seeds.

4.3.3. Cross Pollinated Crops Breeding

Breeding cross-pollinated species would include few most vital methods:

- Mass selection
- Developing hybrid varieties
- Developing synthetic varieties

The purpose of each such methods of breeding would be preservation or restoration of heterozygosity. It is due to the fact that for several traits, the cross-pollinated species would be inherently hybrid (heterozygous) losing vigor while developing into purebred (homozygous).

Few examples of cross-pollinated crops would include alfalfa, rye, herbage grasses, forage legumes, red clover, few varieties of maize, perennial ryegrass, sugar beet and palm oil. New cultivars selection related to cross-pollinated crop species would be a method for changing the frequency of gene associated with desirable alleles in a mixed genotypes population in an attempt to preserve a high degree of heterozygosity.

Hence, it would be the population properties that would be considered important excluding the individual genotypes (like, in self-pollinated crops). Rather than generating a cultivar for the purpose of release, which would be a uniform genotype, the population would be consisting of a complex blend of genotypes, providing desired performance together. Outbreeding cultivars would be generally of two different types which would be established by their maintenance and multiplication methods:

- Open-pollinating population
- Synthetic cultivars

Mass Selection

Based on similar underlying philosophy and assumptions as in bulk method for inbreeding species, mass selection would operate. The breeding scheme being quite simple, would be using natural environmental conditions for changing the frequency of the genotype related to an open-pollinated population.

Through cross-pollinating two different prevailing open-pollinating population, creation of a new population would be done. A representative set (any one plant would not be totally representative of the population) of individuals belonging to each population, would be considered as crossed in such cases.

The seeds would be grown under field conditions for a multitude of seasons, are supposed to have been resulted from the above mentioned set of crosses. An assumption related to random crossing could be made.

It would be resulting into a quickly moving population towards equilibrium considered to be maintained in the form of an exploitable population. It would be worth noting, while isolating such growing population from other crops in relation to the particular species; care would be required, creating growth possibilities within pollination distance.

Hybrid Varieties

A remarkable example related to the hybrid vigor exploitation using varieties of F_1 hybrid would be with corn (maize). While producing a hybrid variety of corn, three steps would be involved:

- Superior plants selection
- Several generations of selfing for producing inbred lines series, which however, would be differing form each other, would be all pure-breeding showing high uniformity
- Inbred lines would be selected through crossing.

The vigor of the lines would dramatically decrease during the process of inbreeding, generally to a less than half compared to field-pollinated varieties. Nevertheless, the crossing of any two unrelated inbred lines would be restoring vigor and the F_1 hybrids among inbred lines would be considerably superior against open-pollinated varieties.

The hybrid among any two inbred invariably remaining the same would be a vital effect pertaining to the homozygosity of the inbred lines. Producing hybrid seeds of any required quantity would be possible once the inbred giving the superior hybrids would be identified.

Corn (maize) pollination would be through wind, blowing pollen from the tassels to the styles (silks) that would bulge from above the ears. Thus, the seed parent interplanted in two or three rows, one row belonging to the pollinator inbred, detasseling the former, prior to its pollens being shed, would economically accomplish regulated cross-pollination on a field scale.

Figure: 4.6: Corn Tassel (pollens would be blown from tassels to styles).

Source: Image by Wikimedia Commons

Effectively, production of a majority of hybrid corn would be from "double crosses", wherein, four inbred lines would be initially crossed in pairs (A×B and C×D) followed by repeat crossing of the two F_1 hybrids (A×B) ×(C×D). The procedure involving double cross would have the benefit of producing the commercial F_1 seed on the single cross A×B which is considered to be extremely productive, instead of a low-yielding inbred, thus, resulting in reduced seed costs.

Elimination of the seed parent detasseling resulted through cytoplasmic male sterility in the recent years. Hence, it led to the production of hybrid seeds in more economies. In the following generation an exceedingly hybrid vigor displayed by F_1 varieties of hybrid is supposed to be disappearing.

As a result, for the purpose of stock plantation, hybrid variety seeds would not be used. However, every year farmers would be buying new seeds from seed companies.

For expanding food supply quantity available to the global population, probably there would not have been any other development of biological sciences, impacting so greatly than the development of hybrid corn (maize). By using male sterility producing hybrid varieties in other crops would be possible and would be hugely successful. It would also signify the usage of hybrid varieties to be continued exponentially in future.

Synthetic varieties

'Synthetic variety' terminology is supposed to have been emerged to be used for designating a variety which would be maintained from seeds of open-pollination, subsequently synthesized by hybridization collectively among several selected genotypes combinations, considered to have been examined related to combining ability.

The synthetic variety components would very likely be inbred (usually), clones, population of mass selection or many different materials. For the purpose of reconstituting the synthetic consistently, the component units would be maintained.

The selected inbred for component lines would be done according to the combining ability tests. In all probable combinations the component inbred would be crossed. The seed that would be inter-crossed is supposed to be known as Syn 0.

From all crosses same quantity of seeds would be composited allowing the mixture to be open-pollinated separately along with the harvesting of seeds. It would be the Syn 1 generation. The population would be developing into an open-pollinated variety without a synthetic reconstitution consistently.

The combing ability of testing would be a conclusive measure in relation to a synthetic variety for distinguishing it from a conventional variety related to a species of cross-pollination emerging in an incessant selection of individuals and following progeny tests.

The occurrence of huge variability through crossing many components with an ability of high general combining would result in the formation of more adaptable synthetic varieties in comparison to conventional varieties.

Much like the hybrids, the synthetic variety yield would usually decrease post the Syn 2, till the time an equilibrium would be touched, in partly self-fertile species. It would be dependent on the rate of selfing and inbreeding (lowest depression), but also on the quantity of components used in Syn 0.

4.4. CLONAL BREEDING

As per a clonally propagated crop definition the material for cultivating and maintain a variety would be gained through asexual reproduction. It would be irrespective of the fact that the plant material being used for propagating is considered to be different within and between species, surrounding tubers, roots, stem cuttings and corms along with seeds developed asexually (seeds developed in absence of meiosis).

It is supposed to be observed that if crops like maize (bred like an open-pollinated or hybrid crop) or beans (bred like a cross-fertilized, self-fertilized or hybrid crop), would supposed to have been propagated by stem cutting or seeds developed asexually.

The crops would be considered as clonally propagated. On the other hand, during clonally propagated crop breeding, the techniques, and methods of breeding, generally in association with cross-fertilized and hybrid crops could be quite beneficial.

For example, parents' selection in potato, cassava and sweet potato breeding would be recombined in polocrosse nurseries of open-pollination for the development of new genetic variation. For the future of clone breeding, the techniques and methods derived from cross-fertilized and hybrid crop breeding would certainly become far more crucial.

Breaking the usual clonal propagation through a crossing step, and thus, resulting in the development of sexual seeds and genetic variation for selecting new clones, would seem to be simple. Hence, the finally selected clone is considered to be genetically same including the original seed plant wherefrom the selected clone is supposed to have been derived.

To rephrase it, every individual seed plant would be considered as an effective variety. Humans are supposed to have been using roots and tubers, fruit and tree plant species, way ahead of the onset of agriculture.

IPB domestication (Simmonds, 1979) as well as a considerable progress in yield through FPB would have been made in few regions of the world. However, there would not be sufficient yield progress in other regions of the world and a strong requirement of PPB for progress would appear in such places.

4.5. NEW GENETIC APPROACHES

4.5.1. Tissue Culture (in vitro) Techniques

Under the name of tissue culture multiple techniques (micro propagation, haploid production, protoplasts culture, embryo culture, apical culture, somatic embryogenesis, etc.) are considered to have been developing. The examples provided would provide an understanding of the probable applications.

4.5.2. Haploidy

An important part in the development of new cultivars in several species of crop would be through the establishment of true breeding, homozygous lines. The usage of the homozygous lines would be in the form of cultivars in their own rights (i.e., for the purpose of inbreeding crop species) or as parents for developing the hybrid variety.

For achieving homozygosity, the process of selfing or mating among near relatives would traditionally be used by plant breeders. Therefore, it is supposed to have been the aim of various plant breeders to get a scope of producing plants from gametic, haploid cells. After the haploid chromosomes get doubled this technique would result in producing inbred lines instantly.

The forming of haploid gametes through meiosis would be resulting from the genetic phenomenon that would be crucial to gain homozygous. The number of chromosomes would be halved at the time of such cell division representing each chromosome for only one time in each cell (with the assumption of the species being basically a diploid one).

Inducing such gametic, haploid cells could result in the development of plantlets (i.e., sporophyte development could be encouraged – note: lower plants usually have a definite life cycle phase) leading to the developing of a haploid plant. Further it could be treated for boosting its chromosomes to double for a totally homozygous line (a doubled haploid) production.

- Techniques used for producing haploids in vitro

Haploid plants occurring naturally would be quite rare for several plant breeders, even though haploidy would be considered as an appealing technique. Nevertheless, by using plant tissue culture, plant production was allowed out of gametic cells culturing in vitro at a greater frequency.

Although it is considered that regenerating haploid plants would be possible from sex cells belonging to both male and female, however, in regenerating huge numbers of haploid and double haploid lines, it would normally be the male cells (microspores or pollen) which are proved to be majorly successful. Partially it would be due to the easy collection of pollens, as against eggs, and partially due to the fact that more pollen grains would be produced in comparison to eggs.

There would be several haploid induction methods which would not have any direct relation with tissue culture, however, the most extensively used would be through anther or microspore (immature pollen grains) culture in vitro.

- In vitro multiplication

There could be two major benefits (specifically in clonal species) associated with in vitro multiplication of breeding lines.

- In vitro propagated plants could usually be originated and preserved in disease-free condition, which could be helpful in maintaining breeding line stocks, facilitating the storage of germplasm for long term as well as facilitating exchange of material internationally.
- The number of plants increasing rapidly could be derived easily from short generation time and vigorous growth.

The above mentioned points would have specific significance for clonal crops. It is due to the fact that clonal crops would likely have a lower rate of multiplication due to their vegetative form of propagation. Moreover, they would be specifically vulnerable towards viral and bacterial diseases which would be prone for multiplication and transmission by means of each clonal generation.

Potato and strawberry would be good examples for the maintenance of disease-free status along with providing potential to regenerate plant rapidly. Other examples, probably related to the less matured ones, would be including date and oil palms in vitro propagation.

It was established that in such crops plant regeneration in a more rapid manner would certainly provide a substitute for the slow and prolonged propagating process of side shoots in date palm as well as an additionally consistent material for planting in oil palm.

Nevertheless, the process would be anyway be too much dependent on genotype. In case of palm oil, it would be considered that initially there would be a proof of producing an unsuitably high frequency of sterile palms.

4.6. MUTATION BREEDING

In self-pollinated crops which are supposed to be disseminated by seed, mutation breeding would be in accordance to self-fertilization – or selfing – of mutants till the time the desired character induced would be expressed stably in generations of advanced mutants. Often it would be essential to backcross to the original non-mutated genotype (part of the DNA series related to a cell determining its particular characteristic) for the retention of its favorable characteristics.

Mutation breeding would be based on mutation induction and mutation detection. It has several relative advantages: it is economical, fast, confirmed, and vigorous. It would also be transferrable, universally applicable, safe, and environmentally friendly.

Mutant varieties exceeding 3,200 which would include various crops, ornamentals and trees, are considered to have been released officially for the purpose of commercial use in over 210 species of plants from more than 70 countries (Source: FAO/IAEA Mutant Varieties Database).

In mutation breeding, plant biotechnologies would be playing a major part. For minimizing the required time for the generation of mutant lines breeding (the ones that would always transfer few traits to their offspring), techniques of plant tissue culture are considered to be effective tools.

For exploitation of the induced crop mutation being regarded as recessive (in genetics, when one feature of a gene would not be expressed due to the displaying of a more dominant one), there would be a bottleneck.

The double haploid technique would be one of the breeding tools referring to the chromosomes of a haploid being doubled – with a single member of each chromosome pair, belonging to an organism or a cell. Another method is supposed to be the identification of molecular marker that would be associated closely to the desired traits, and which would be able to identify such traits rapidly.

To additionally strengthen the plant mutation breeding programs, developing, and disseminating of the above mentioned molecular markers are considered to be potential. It would be specifically for major food crops, like rice. In the genetic material pertaining to living organisms, mutation would be a heritable transformation. Hence, it would be considered as a chief cause for the diversity and evolution of species.

Figure: 4.7: Plant Breeding Center.

Source: Image by Flickr

Mutations are believed to be spontaneous or induced. Mutants which would be more suited to their environment (natural or man-made) would have a selective benefit. For bringing wild species into domestication and agriculture mutation is considered to have been a great factor.

Pre-conceived mutation is considered to have been a highly effective strategy in plant breeding. Presently more than 3,220 mutant cultivars are supposed to have been released officially in more than 210 plant species. Producing mutant traits for the majority would be possible, though excluding a few breeding goals.

It would be inclusive of yield, quality, stature, disease, resistance towards pest, abiotic stress tolerant, and degradation after harvesting and new end-user characters. Presently, the huge majority of mutations which would be used in breeding of plants and producing crops would be originating from physical mutagen treatments.

4.6.1. Spontaneous Mutation

Spontaneous mutations would be the naturally occurring mutations. They would be regarded as both germinal and somatic. Solely in vegetatively propagated plants, incorporation of valuable somatic mutations would be applied in crop improvement, like seedless grape, naval orange, Bhaskara banana.

Vegetative propagation would also be considered beneficial for the maintenance of germinal variation achieved by means of sexual reproduction, like apple, mango, potato, sugarcane. Hence, the origin of every genetic variations happening in all living things presently, would be from spontaneous mutation.

4.6.2. Mutagens and Induced Mutations

The frequency of spontaneous mutations is considered to be very low. Hence, the mutation frequency would be raised through few agents known as mutagens. There would be two types of mutagens:

- Chemical mutagens
- Physical mutagens.

Few chemicals, like ethyl methane sulphonate (EMS) and sodium azide, consist of chemical mutagens inducing mutagens.

Various types of radiations like X-rays, gamma-rays, ultraviolet rays, etc., contain physical mutagens causing mutations. Changes in DNA and chromosomes would be induced by such mutagens, producing mutations. While responding to mutagens the mutations produced would be termed as induced mutations. As per records, they are considered to have been produced first by Muller (1927) by using X-rays on Drosophila and by Stadler in case of maize. Therefore, mutation breeding would be explained as using induced mutations in the breeding of plants for the development of improvised varieties.

4.7. GENETIC ENGINEERING (RECOMBINANT DNA TECHNOLOGY)

It is regarding a process wherein; deliberate and artificial means would be used for genetic makeup of cell alterations. Transferring and replacing of genes for recombinant DNA creation would be included in the process. It would be executed by the DNA molecules being cut at particular sites for obtaining fragments with desirable and beneficial genes from one type of cell. Henceforth, such genes would be inserted within an appropriate carrier or vector.

It would be followed by putting the recombinant DNA within a totally different cell belonging to a bacterium or plant or animal cell. They would be acquiring valuable characters through the above mentioned method, like resistance towards diseases or making of valuable enzymes, hormones, vaccines, etc.

The DNA (genes) manipulation or engineering would alter the attributes of the above mentioned process, hence, being named as 'genetic engineering'. Cloning and amplification of the recombinant DNA molecules would be done to an unrestricted extent.

4.8. POLYPLOIDY BREEDING

Polyploid is considered as an organism with over two chromosome sets or genomes per cell. The particular state would be called polyploidy. The major important crops like wheat, bananas, cotton, potatoes, sugarcane, and tobacco are supposed to contain polyploidy condition.

The occurrence of polyploidy in nature would be due to chromosomal failure for getting detached during anaphase. Non-disjunction or non-formation of spindle would be the reason behind it during mitosis or meiosis. By applying colchicine there would be a probability to artificially induce it.

The number of genomes in a polyploid would decide whether it would be called a triploid (3n), tetraploid (4n), pentaploid (5n), hexaploidy (6n), etc. Odd number of genomes in polyploids (triploids, pentaploid) would be considered as sexually sterile due to the non-formation of synapsis in odd chromosomes.

Therefore, they would be vegetatively propagated. For example, banana, pineapple. Moreover, cross-breeding with diploids in a free manner would be absent in polyploids. There would be two types of polyploidy – auto polyploidy and allopolyploidy.

4.8.1. Auto polyploidy

A polyploidy with an increasing number of genomes of the same type would be known as auto polyploidy. For example, auto triploid (AAA), auto tetraploid (AAAA). Auto polyploids would be in few of the crops as well as garden plants, like maize, rice and gram. The gigas effect would be induced by auto polyploidy.

4.8.2. Allopolyploidy

Hybridization among two species following the chromosomes getting doubled is considered to have resulted in developing the allopolyploidy (e.g., AABB). The common type would be the allotetraploid. Allopolyploids are supposed to be operating in the form of new species, e.g., wheat, American cotton and *Nicotiana tabacum* etc.

Two of the allopolyploids produced recently would be *Raphano brassica* and Triticale. Hence, Triticale would be considered as a wheat hybrid (*Triticum turgidum*) and rye hybrid (Secale cereale). Triticale would be the first man-made crop produced as a result of wheat and rye crossing.

Autoallopolyploidy is considered to be a kind of Allopolyploidy wherein a single genome would be in an exceeding diploid state. Usually autoallopolyploids would be hexaploids (AAAABB). For example, *Helianthus tuberosus*.

4.9. CONCLUSION

In agricultural biotechnology, the objective would be towards plants and livestock's improvisation, which would include separation of genes from an organism, modifying and manipulating the genes and restoring them back into several organisms. Though, plant breeding would have been initiated in the earlier times, the procedures and technologies would be enhancing with time.

In addition to using pollination in the form of Mendelian crossing method, for crop breeding, there would be several other techniques, invented, at present for plant breeding. The most developed form of techniques used for breeding plants would be plant tissue culture and propagation, marker supported selection and genetic engineering.

For improving the germplasm, various methods of breeding would be available to plant breeders. For example, the pedigree method, bulk method, single seed descent, backcross method and recurrent selection. A modified pedigree program and to a reduced level of backcross, single seed descent, recurrent selection and double haploid breeding methods would be mainly used in the North Dakota breeding program.

Therefore, plant breeding would be considered as a science that would be dependent on the principles and practices of genetics and cytogenetic, i.e., it would be called applied plant genetics. It would be emphasizing on the genetic makeup improvement of the plants for the development of enhanced varieties. In other words, for the production of more favorable products it would be manipulation of plant genetics.

Plant breeding is supposed to be occurrence of selection through the presence of variability. In short, it is manipulating biological system necessitating several generations for achieving dynamic, exciting, and challenging professing functioning under repeatedly changing conditions.

REFERENCE

1. 2015. Plant Breeding Reviews: Volume 39. [online] 39. Available at: <https://onlinelibrary.wiley.com/doi/abs/10.1002/9781119107743.ch2> [Accessed 13 June 2020].
2. Agritech.tnau.ac.in. n.d. Crop Improvement: Breeding Methods in Crop Plants. [online] Available at: <http://agritech.tnau.ac.in/crop_improvement/crop_imprv_breed.html> [Accessed 13 June 2020].
3. Allard, R., 2019. Plant Breeding - Evaluation of Plants. [online] Encyclopedia Britannica. Available at: <https://www.britannica.com/science/plant-breeding/Evaluation-of-plants> [Accessed 13 June 2020].
4. Biology Discussion. n.d. Crop Improvement Selection Methods (With Diagram). [online] Available at: <https://www.biologydiscussion.com/crops/improvement/crop-improvement-selection-methods-with-diagram/17663> [Accessed 13 June 2020].
5. Biology Discussion. n.d. Plant Breeding: Steps and Methods of Plant Breeding for Disease Resistance! [online] Available at: <https://www.biologydiscussion.com/plants/plant-breeding-steps-and-methods-of-plant-breeding-for-disease-resistance/1340> [Accessed 13 June 2020].
6. Breseghello, F. and Coelho, A., 2013. Traditional and Modern Plant Breeding Methods with Examples in Rice (Oryza sativa L.). Journal of Agricultural and Food Chemistry, [online] 61(35). Available at: <https://pubs.acs.org/doi/pdf/10.1021/jf305531j> [Accessed 13 June 2020].
7. Caligari, P., 2001. Plant Breeding and Crop Improvement. [ebook] Available at: <https://www.researchgate.net/publication/227986002_Plant_Breeding_and_Crop_Improvement> [Accessed 13 June 2020].
8. Grüneberg, W., Mwanga, R., Andrade, M. and Espinoza, J., 2009. Breeding Clonally Propagated Crops. [ebook] Available at: <https://www.researchgate.net/publication/301700183_Breeding_clonally_propagated_crops> [Accessed 13 June 2020].
9. Iaea.org. 2011. Mutation Breeding | IAEA. [online] Available at: <https://www.iaea.org/topics/mutation-breeding> [Accessed 13 June 2020].
10. Breseghello, 2011
11. Castro, 2011
12. Clark, 1997

13. Crops et al., 1992
14. Falconer, 1996
15. Guimaraes, 2011
16. Gumber, 1998
17. Hallauer, 2005
18. Harlan, 1992
19. Hartl, 1997
20. Johnson, 1965
21. Keijer, 2008
22. Kleter, 2008
23. Kok, 2008
24. Kuiper, 2008
25. Lonnquist, 1965
26. Lopes, 2011
27. Mackay, 1996
28. Melchinger, 1998
29. Melchinger, 2005
30. Moll, 1965
31. Morais, 2011
32. Muller, 1927
33. Oliveira, 2011
34. Pereira, 2011
35. Pinheiro, 2011
36. Reif, 2005
37. Shull, 1948
38. Silva, 2011
39. Simmonds, 1979
40. Utumi, 2011
41. Velez-Fortuno, 1965

5

Types of Mating System

CONTENTS

This chapter is about the mating system which basically means that a group which is being structured in the relation to the sexual behavior. This chapter gives the brief description about mating system used in crops and types of mating system involved.

This chapter gives a brief description about the importance of mating system and how this system came into working. This also involves the concept behind the mating system in the self-compatible hermaphrodite. This chapter is also giving a brief idea about the importance of mating system with the future prospective for studying the mating system.

5.1. INTRODUCTION

Diversity is said to be the rule in living organisms. Whereas, on the other hand, this diversity is manifest at a number of different levels of the life tree; in the vegetable kingdom, the diversity is possibly considered to be the most apparent form. This was revealed by the high diversity of plant morphologies and the life histories even at small spatial scale.

In plant biology, when the first investigations was done, then the botanists have always focused on the high variation of reproductive systems considering the plants and also, that the floral diversity (forms and colors) in higher plants is considered to be one of the most obvious forms of variation.

A basis was provided from that in order to discriminate and classify the plants. In the period of 18th century, there has been variation in sexual structures of plants. Therefore, this has provided the basis for the Linnaean classification. Interestingly, there was variation which in turn reveals variation as well as adaptation of the mating system and furthermore, it results from evolutionary processes in the phylogeny.

Furthermore, the mating systems are said to be the central in population biology first because of the fact that it tends to ensure the maintenance (and ultimately the growth) of populations. The second fact is that because it tends to shape the transmission of phenotypic traits through the transmission of the hereditary material which in turn conditions the evolutionary processes.

From a long time, if the diversity of plant reproductive systems as well as floral morphologies have intrigued the naturalists. There have been botanic studies which have long been only descriptive, and also, do not consider any evolutionary interpretation for the rise of such diversity. Darwin was the first who proposed the first evolutionary interpretation. Darwin devoted three volumes on plants reproductive biology (Darwin, 1877).

The processes of pollination and the dependence to pollen vectors were considered to be the central selective force in view of Darwin. The rise of Mendelian Laws as well as more recently population genetics, specifically the work of Sir Ronald Fisher in the period of 1940s, have laid the foundation for a solid theoretical framework. This framework was based on gene dynamics. Contrastingly, there has been development of the botanical tradition which took place in a more empirical way.

These two were the historical traditions that have given birth to two different approaches. These different approaches have remained relatively separated until recently (Uyenoyama et al.1993). Considering the last ten years, the rapprochement is perceptible (Barrett, 2008). Interestingly, the studies related to the mating system in plants is said to be the good example of fruit-full interaction between field data, theory, and experiments.

Field observations of flowering plants, interactions with pollinators are considered to be an important corpus of data. In addition to this, the mating system theory is said to be an active field of research which addresses the major issues in evolutionary biology including kin selection the effect of deleterious mutations or mutual interactions.

Ultimately, plant mating system is said to be an area where the experimental approach in order to test particular hypotheses which has been successful thanks to suitable tools as well as techniques.

Let's take an example. The self-fertilization can be measured precisely under natural conditions and this was done by genetic markers. This can also be manipulated in laboratory which in turn allows testing adaptive hypotheses.

Generally, the mating systems are defined as the mating patterns among individuals which majorly refer to the selfing or outcrossing rates and gene flow.

The mating systems are considered to be the important part of plant breeding systems which in turn control genetic exchange. Also, these play a key role while determining the genetic structure of plant populations, along with this; they also have a profound influence on survival as well as evolution of plant species.

Along with this, the background knowledge of mating systems tends to help in developing the genetic improvement strategies and also, in establishing the relevant conservation measures.

Woody bamboos (Poaceae: Bambusoideae) contain ca. Eighty genera which include more than 1200 species, that are majorly distributed in the subtropical as well as tropical regions of Asia, America, and Africa. Woody bamboos are said to be shrub forest species and this have important economic value including food, construction, and pulp materials, and also, important roles in water and soil conservation.

There is one of the most compelling biological properties of woody bamboos which majorly tend to concern their unusual flowering phenology. There are two common unofficial terms which are related to flowering as well as mating systems of bamboos. First and foremost, this was based on origin of woody bamboo forests, and generally, the bamboo species can be divided into wild as well as cultivated species.

The second point is that in terms of the size of flowering forests or populations, the woody bamboo flowering events can further be divided into two types on the whole, which are named mass and sporadic flowering.

Because of the outstandingly long vegetative growth periods (ca. 20–150 years) among seed plant and uncertainty of flowering events, and so far, much little is known in relation with the mating systems of most bamboo species. This is greatly hindering studies on their germplasm conservation as well as genetic improvement.

Considering the characteristics of mating systems, a number of the documented woody bamboos were dichogamous as well as protogynous, and these were generally said to be wind - pollinated species. Nevertheless, it is still in controversy related to whether bamboos are obligate outcrossers.

Considering the tropical bamboos, the *Dendrocalamus* spp. tendsto demonstrate a breeding system which is suitable for hybridization. In recent times, the paternity analyses of a number of wild temperate woody bamboo species, like Sasacernua (sporadic-flowering), *S. veitchiivar*. hirsuta (mass-flowering) and *Phyllostachys edulis*(sporadic-flowering) were reported.

There are a number of studies which were based on results from single bamboo flowering events. It seemed that the species that studied were self-compatible. At the same time, it appeared that selfing was said to be predominant in sporadic flowering events whereas the outcrossing was predominant in mass flowering events.

5.2. DARWIN ON MATING SYSTEMS

There were contributions made by Darwin which provided two separate discussions on mating systems. The first discussion was concerned with the animal mating systems. In this, the sexual selection occurs further confirming his understanding that the sexual differences arose within this context.

The second contribution made by Darwin on mating systems was concerned with those of plants. In the Different Forms of Flowers on Plants of the Same Species, Darwin provided a detailed explanation about the various contrivances through which plants tend to encourage or discourage cross pollination.

This was the statement which tends to confirm that the understanding of Darwin about the selfing was non-ubiquitous among plants, and also, that the floral differences arose within this context.

Similar to most of his contributions, original emphases provided by Darwin have determined the research avenues which are being pursued in today's scenario. There are two distinct emphases in current mating system research which are almost certainly the result of distinction between plants and animals provided by Darwin.

The emphasis developed majorly for plants as well as domestic breeding designs are expressed in terms of the genetic relationships between males as well as females. The emphasis developed majorly for animals, and this followed the lead of Darwin. This is expressed in terms of mate numbers per male or per female.

5.3. PLANT MATING SYSTEMS

There was an approach provided by Darwin (1877) to plant mating systems which was majorly classificatory. Furthermore, he cataloged floral morphology as per to its rich existing lexicon, and only secondarily developed the hypotheses for the question 'why this variation might have evolved'. Nevertheless, the catalog seems intended to have been provocative provided by Darwin.

Further, he described species with exclusive selfing, and along with this, he considered in careful, parallel detail the mechanisms through which selfing was prevented in specific species. Wright (1922) and Fisher (1941) later explored the population genetic consequences of different levels of selfing as well as outcrossing. Wright (1922) further showed how selfing increased allelic auto zygosity, and therefore, resulted in inbreeding depression most of the time.

Fisher (1941) showed the way in which despite inbreeding's potentially deleterious effects, selfing could increase in frequency while the selfers contributed selfed along with the outcrossed seeds to each generation.

There has been recent research on plant mating systems which has focused primarily on deviations from random mating (1996) in consistent with these early emphases. Along with increasing resolution of genetic markers, the population genetic approaches to plant mating systems and furthermore, their theoretical foundations have become increasingly sophisticated.

Selfing, self-incompatibility, dioecy and gynodioecy tend to appear to have evolved by remarkably simple means (1978-2006). The obligate selfing and obligate outcrossing tend to appear in order to persist as alternative stable states (Lande R, Schemske D, 1985).

While, the mixed mating systems seem constrained by the deleterious effects of inbreeding, despite the fact that the frequent assumptions that a continuum can exist (Schemske D, Lande R, 1985).

However, the frequency-dependent and density-dependent processes in pollination transmission appear to have a powerful influence on whether selfing or outcrossing is stable within a given species (Holsinger K, 1991).

This result is said to be consistent along with observations that wind pollinated plants conform to the predictions of Lande and Schemske (Lande R, Schemske D, 1985; Schemske D, Lande R, 1985), while the animal pollinated plants do not (Vogler D, Kalisz S, 2001).

Moreover, the patterns of floral morphology correspond the relative amount of pollen available in order to export as well as to the density of conspecifics (Barrett S, Harder L, 1996).

There have been debates going on the possibility that the sexual selection can operate within plants (Shuster S, Wade M, 2003) which has further been identified in three primary contexts in which sexual selection might occur:

- differential pollen transfer,
- differential pollen tube growth, and
- maternal control of seed set.

Nevertheless, despite the fact that the male as well as female fitness can vary within each of these contexts in plants, the extreme variance in male fitness which can be seen in animals tends to appear to be constrained by limited opportunities for pollen dispersal, by large within-individual variation in seed size and seed set or by inconsistent success of particular pollen tubes among different stigmas, which is due to pollen limitation most of the time (Ashman T-L, et al. 2004; Knight T, et al. 2005).

Despite it seem to be possible that male-male competition and female preferences in plants, they do not result in a sex difference in fitness variance which can be compared in magnitude to that of animals. Therefore, although there is rigor of their theoretical foundations and further, hints of similarities in mechanisms, there have been studies of plant mating systems which have remained conceptually divorced from those of animals.

5.4. TYPES OF MATING SYSTEM

In plants, the primary mating systems are outcrossing (cross-fertilisation), autogamy (self-fertilisation) and apomixis (asexual reproduction without fertilization and this tend to only take place when arising by modification of sexual function). In plants, in mixed mating systems, the plants use two or even all three mating systems, are common.

5.4.1. Cross Fertilization

Cross-fertilization is referred to as Allogamy. The cross fertilization is said to be the fusion of male and female gametes (sex cells) from a number of different individuals of the same species. The cross-fertilization is required to occur in dioeciously plants (these are those which have male and female organs on separate individuals) and also, in all animal species in which there are separate male as well as female individuals.

Considering the hermaphrodites (those organisms in which the same individual tends to produce both sperm as well as eggs), a number of species possess well-developed mechanisms which make sure that the process of cross-fertilization occurs. Furthermore, a number of different hermaphroditic species which are capable of self-fertilization (q.v.) also have capabilities for the purpose of cross-fertilization.

There are a number of different ways in which the sex cells of two separate individuals can be brought together. Considering the case of lower plants, such as mosses and liverworts, it is said that the motile sperm are released from one individual and furthermore, swim through a film of moisture to the egg-bearing structure of another individual.

Figure 5.1: Cross pollination done by fly.

Source: Image by Flickr

In the case of higher plants, the process of cross-fertilization is achieved through cross-pollination, when pollen grains give rise to sperm and these are further transferred from the cones or flowers of one plant to egg-bearing cones or flowers of another.

The occurrence of the cross-pollination can take place by wind, as in conifers, or through the symbiotic relationships with a number of different animals (for example, bees and specific birds and bats) which further carry pollen from plant to plant whereas the feeding on nectar.

There are a number of methods of cross-fertilization which are equally diverse in animals. There are a number of species which breed in aquatic habitats, the males as well as females each tend to shed their sex cells into the water and also, the external fertilization takes place.

However, considering the terrestrial breeders, fertilization is internal, along with the sperm which are being introduced into the body of the female. The internal fertilization also tends to occur among some fishes and some other aquatic breeders.

The cross-fertilization helps maintain a greater range of variability for natural selection in order to act upon, thereby increasing the capacity to adapt to environmental change of the species, which is done by further recombining genetic material from two parents.

5.4.2. Self-Fertilization

In the process of self-fertilization, the fusion of male and female gametes (sex cells) produced by the same individual occurs. The process of self-fertilization tends to occur in bisexual organisms, and these include a number of flowering plants, numerous protozoans, and a number of invertebrates.

Autogamy is said to be the production of gametes by the division of a single parent cell, and this is frequently found in unicellular organisms like the protozoan *Paramecium*.

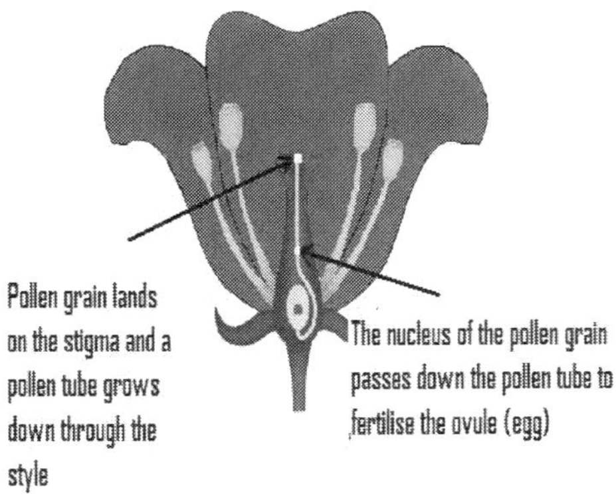

Figure 5.2: Image representing self-fertilization.

Source: Image by Wikimedia Commons

However, these organisms can also do reproduction by means of conjugation. In this, the cross-fertilization is achieved through exchanging the genetic material across a cytoplasmic bridge between two individuals.

Similarly, among higher plants, a number of these are monoecious (this means that they are bisexual that is the male as well as female gametes being produced by the same individual). There are a number of self-pollinating species which are also capable of cross-fertilization and even those which are obligate self-fertilizers are occasionally cross-pollinated by accident.

There are hermaphroditic animals in which both male and female gonads are borne on one individual. These are rarely capable of self-fertilization because of the fact that there are a number of such species which have adaptations in turn encouraging cross-fertilization.

Considering an evolutionary and reproductive mechanism, the self-fertilization tends to allow an isolated individual in order to create a local population and moreover, it stabilizes desirable genetic strains. But it fails in order to provide a considerable degree of variability within a population and further, it limits the possibilities for adaptation to environmental change.

5.4.3. Apoximis

In flowering plants, apoximis is defined as the asexual formation of a seed from the maternal tissues of the ovule, while avoiding the processes of meiosis and fertilization. This in turn results in development of embryo.

In higher plants, the initial discovery of apomixis is attributed to the observation which a solitary female plant of *Alchornea ilicifolia* (syn. *Caelebogyne ilicifolia*) from Australia continued to form seeds when it was planted at Kew Gardens in England (Smith, 1841).

The term apomixis which mean "*substitution of sexual reproduction by an asexual multiplication process without nucleus and cell fusion*" was introduced by Winkler (1908). Thus, there have been some authors who have chosen to use apomixis in order to describe all forms of asexual reproduction in plants, but this wider interpretation is no longer generally accepted.

The current usage of apomixis is synonymous along with the term "agamospermous" (Richards, 1997). The reason for this is that the seeds are found only among angiosperm as well as gymnosperm taxa; furthermore, this definition of apomixis tends to limit its use to those groups. In the case of lower plants, the phenomenon is same as the apomixis which is known.

But there have been discussions which remain about the use of this term in cases in which the reproductive structures involved are different, but still they are considered analogous (Asker and Jerling, 1992).

Despite the fact that it is sometimes referred to as a botanical curiosity, and so, the apomixis is far from rare, this is being relatively prevalent among angiosperms, along with a pattern of distribution. This distribution suggests that it has evolved a number of times.

It has been described in more than 400 flowering plant taxa, and these include representatives of more than 40 families (Carman, 1997). It is well represented among both monocotyledonous as well as eudicotyledonous plants; curiously, though, it tends to appear to be absent among the gymnosperms.

This provides further estimation which is almost certainly very conservative. Unequivocal confirmation of apomixis further requires the simultaneous examination of both genetic as well as cytological evidence (Nogler, 1984).

The embryological examination of plant taxa for apomixis has not been exhaustive, and furthermore, provides support genetic evidence which is uncommon even in the case, when the apomixis has been declared for a particular plant.

It seems likely that as there is increase in the understanding of this phenomenon and methods in order to determine its presence improve, there are a number of more angiosperm taxa which will be found to include apomictic representatives and some other suspected cases will be revised. There has been a recent study done by Plitman (2002) which supports this prediction.

There have been a number of several authors which have noted a marked bias in the distribution of apomixis among angiosperms (Asker and Jerling, 1992; Mogie, 1992; Carman, 1997; Richards, 1997). There are plants which are known to use gametophytic apomixis, around 75 percent of confirmed examples belong to three families, which are named as the Asteraceae, Rosaceae, and Poaceae. These collectively constitute only 10 percent of flowering plant species.

In converse to this, despite the fact that the apomixis is known among the Orchidaceae which is the largest flowering plant family, it tends to appear to be comparatively uncommon among these plants.

There are a number of some authors which have postulated that the current patterns of distribution can reflect the predisposition of specific plant groups to the unique developmental as well as genetic changes which further characterize apomixis (Grimanelli et al., 2001).

This is the hypothesis which appears intuitively attractive, but on the other hand, this (like many issues related to the apomixes) remains conjecture until unless, it is tested experimentally. Some of this bias also can relate to the ease of embryological examination in some of the plant groups or to data which is accumulated from embryological investigations related to the activities in crop improvement.

There are some other noted associations between apomixis and a number of different plant life history traits which in turn provide insight into the nature and possible ecological role of this phenomenon. The apomixis frequently is related to the expression of mechanisms which tend to limit self-fertilization (autogamy).

A number of apomictic plants belong to genera and in this the sexual members predominantly tend to exhibit physiological self-incompatibility, dioecy, or heterostyly (Asker and Jerling, 1992). Considering some cases, it can be said that it is clear that the apomicts themselves have retained such mechanisms.

The dioecy is known in a number of apomicts, like *Antennaria* (O'Connell and Eckert, 1999), Cortaderia (Philipson, 1978), and Coprosma (Heenan et al., 2002).

In the similar manner, the self-incompatibility is said to be a known characteristic of the sexual biotypes of *Hieracium* subgenus *Pilosella* (Gadella, 1984, 1991; Krahulcovà et al., 1999). This was demonstrated in recent times in the apomictic species viz. *H. aurantiacum* and *H. piloselloides* (Bicknell et al., 2003).

In addition to this, the apomictic species are almost invariably perennials. Most of the time, they use a vegetative mechanism of asexual reproduction like the stolon or rhizome growth. Therefore, in the field, the apomicts can form large clonal stands through a combination of apomixis and vegetative division, and these can persist through long periods of time.

Also, the apomixis frequently results in the formation as well as maintenance of a number of morphologically distinct, but still interfertile, varieties growing true to type from seed. The taxonomy of such agamic complexes can be a difficult as well as contentious task (Dickinson, 1998; Horandl, 1998).

There can be examples of genera in which the apomictic mode of reproduction is considered to be strongly combined along with the morphological polymorphism. These include *Alchemilla, Potentilla, Hieracium, Poa, Rubus, Ranunculus, and Taraxacum* (Czapik, 1994).

The apomicts are commonly found in habitats which are frequently disturbed or the habitats where the growing season is short, these can include arctic and alpine sites. Their habitat can also be a place where other barriers tend to operate in order to inhibit the successful crossing of compatible individuals, including among widely dispersed individuals within a tropical rain forest (Asker and Jerling, 1992).

Along with this, there are clear indications that the glaciation events tend to define the distribution patterns of a number of agamic complexes (Asker and Jerling, 1992). Gametophytic apomixis is considered among herbaceous and tree species, but it can be said that it is considerably more common among the former.

This can be simply a reflection of the predominance of the trait in the Poaceae as well as the Asteraceae. Both of these contains largely of herbaceous species. In the similar manner, the gametophytic apomixis is common among plants which have dehiscent fruits, which can also be seen in the Poaceae, Asteraceae, and Ranunculaceae. But again, this can be more coincidence than a causal connection.

5.5. EVOLUTION OF MATING STRATEGIES

When one considers that the sole function of the structural variation in flowers and inflorescences and different mechanisms is to promote the reproductive success, then it is considered to be for deployment of male and female gametes which is remarkable.

The analysis of any plant community tends to reveal a number of various pollination as well as mating systems which coexist under apparently similar ecological conditions.

The close relatives can reproduce in a number of different ways just as unrelated taxa most of the time tends to share similar floral adaptations. The question thus arises that 'how can biologists hope to make sense of this variation as well as the evolutionary processes which are responsible for the impressive reproductive versatility of flowering plants?'

The plant sexual diversity can be completely understood and for this it is important for biologists to determine the physiological, developmental, and molecular mechanisms which tend to govern the way in which the reproduction occurs. Also, it is important biologists to consider the reason why the evolution of specific reproductive adaptations has taken place. Evolutionary biology provides the tools for solving such kind of problems.

There are some evolutionary studies of plant mating strategies which tend to employ three complementary approaches:

- Theoretical models: these tend to identify key parameters which govern the dynamics of change in mating patterns and also, the use optimality criteria for investigating the evolutionarily stable reproductive strategies;

- Ecological and genetic experiments on populations: these tend to focus on the selective mechanisms which are responsible for the evolution as well as maintenance of sexual traits, and also, the fitness consequences of specific mating patterns, especially considering the degree of inbreeding; and
- Phylogenetic and comparative methods: these tend to explore historical questions which are concerned with the evolutionary origins of reproductive systems. These test adaptive hypotheses which are concerned with functional connections in between traits and mating success.

These approaches are used in order to gather information and this information can be integrated in an effort for showing the way in which the micro evolutionary processes tend to influence the macro evolutionary patterns.

The sexual traits are considered to be important in this regard and the reason for this can be that the genes influence reproduction and these genes govern their own transmission and that of all other genes in an organism. Therefore, the shifts in mating pattern can have profound influences on genetic variation, speciation, and evolutionary diversification.

There has been research on the evolution of plant mating strategies which tend to have two main themes. These themes are related to the evolution of combined versus separate sexes, and cross- versus self-fertilization.

Both of these dichotomies present their own specified challenges. But in each and every case, the main aim is to determine the selective forces which are responsible for the evolution of one condition over another, or responsible for the maintenance of stable mixed strategies which involve dual reproductive modes.

There is an attractive feature of flowering plants and that is a number of taxonomic groups tend to display considerable inter- and intra - specific variation which encompass different reproductive systems.

This type of variation further allows for model building in order to predict conditions which favor shifts in reproductive mode. In addition to this, it also allows for direct testing of these models which can be done on the basis of the ecological and life-history associations which tend to accompany such changes.

5.5.1. The Role of Pollinators in the Mating System

There have been a number of studies related to the mating system patterns which seldom incorporate the dynamics of the pollination process. Thus, the mating system mechanisms are divorced from mating system outcomes most of the time (Harder and Barrett, 1996).

Let's take an example of a few studies which can directly connect pollinator movements to the realized mating system of flowers visited by those pollinators and these include Brunet and Sweet, 2006; Karron et al., 2004; 2009; Karron and Mitchell, 2012.

Along with this, it is important to understand a synthetic approach in order to answer some questions such as:

- How do reductions in pollinator visitation rates and S-allele diversity interact in effects on reproductive success (Young et al., 2012)?
- How would the loss or decline of the primary pollinator of a species influence out-crossing rates, mate diversity and patterns of pollen-mediated gene dispersal?
- How does sharing pollinators with co-occurring plant species affect the mating system (Bell et al., 2005; Moeller, 2006)?
- How does pollinator behavior vary among morphs in plants with sexual polymorphisms?
- How does this variation affect the mating system and the relative success of different sexual types (Ashman, 2000; Case and Ashman, 2009)?

In order to understand and answer such question it is important to go through studies of pollinator visitation patterns and precise genetic estimation of the mating system thoroughly, and thus, it will provide more comprehensive insights into the evolution of plant reproduction.

5.6. CONCEPT OF MATING SYSTEM IN SELF-COMPATIBLE HERMAPHRODITES

When a pollinator visits a hermaphroditic flower then most of the time it deposits a mixture of pollen from a number of different sources, and this include self-pollen, pollen from close relatives, and pollen from one or more unrelated donors (Waser, 1993; Karron et al., 2006).

Despite the fact that this diverse pollen pool is dichotomized as 'self- vs. outcross' most of the time, it is believed that a deeper understanding of the diversity of outcross pollen, the factors which influence these mating patterns, and the fitness consequences of such mating events is required.

There is one important component of a diverse mating pool and that is crossing between close relatives (this is referred to as bi-parental inbreeding). This can be especially likely in populations which have substantial genetic structure (Waller, 1993; Griffin and Eckert, 2003).

When movement of pollinators takes place to short distances and pollen carryover is limited, then a large fraction of the pollen gets deposited on a stigma which can come from close relatives (Mitchell et al., 2009).

Substantially, these inbred (but non-self) mating can decrease heterozygosity and, if there is presence of inbreeding depression, then these lower offspring fitness (Nason and Ellstrand, 1995). This decrease in fitness is because of the biparental inbreeding which is considered to influence the evolution of mating systems and this is done by changing the relative costs as well as benefits of selfing versus outcrossing (Uyenoyama, 1986; Yahara, 1992; Waller, 1993).

Let's take an example. When deleterious recessive alleles are expressed through biparental inbreeding, then there is reduction in the fitness difference between self and outcross progeny (Waller, 1993).

Surprisingly, the rare quantification has been done to the extent of bi-parental inbreeding (b) and the fitness consequences of bi-parental inbreeding depression in natural populations. There have been a number of attempts in order to quantify bi-parental inbreeding by using marker genes and these attempts have compared estimates of selfing on the basis of single marker genes (s_s) with estimates based on multiple marker genes (s_m; Brown, 1990).

Despite the fact that the theory suggests that $s_s - s_m$ will give an estimate of b (Shaw et al., 1981), usually, this parameter is substantially underestimated because of an insufficient number of marker loci (Leclerc-Potvin and Ritland, 1994; Ritland, 2002).

There is a more powerful approach and that is to compare the selfing rate of experimental populations which have been randomly rearranged. This is done in order to break up genetic structure along with the selfing rate of paired control populations where the plants were returned to their original locations (Kelly and Willis, 2002).

This approach is used, and then a study of *Aquilegia canadensis* demonstrated that approximately 30 percent of all mating were between close relatives (Griffin and Eckert, 2003). Provided the high level of inbreeding depression in *A. canadensis*, full sib mating is likely to have a dramatic effect on fitness (Herlihy and Eckert, 2004).

Figure 5.3: Image representing hermaphrodite.

Source: Image by Wikipedia

There was much more work which is needed in order to assess thoroughly the extent of bi-parental inbreeding in natural populations, for quantifying its fitness consequences, and also, for determining the ecological conditions which tend to promote bi-parental inbreeding.

In addition to this, the dichotomy of selfing versus outcrossing masks important differences in the diversity of donors siring seeds within the fruits (Bernasconi et al., 2004). Furthermore, the theory provides suggestions that the clutches sired by multiple fathers will have greater variation in progeny performance.

Furthermore, increasing the likelihood which favorable combinations of maternal as well as paternal genes will be generated (Falconer, 1981; Yasui, 1998; Whittingham and Dunn, 2006).

Also, an increase in mate diversity within fruits tends to decrease genetic relatedness among siblings (Ritland, 1989), moreover, potentially intensifying competition among the maternal half - sibs (Karron and Marshall, 1993; Bernasconi, 2003).

There have been a number of researchers which tend to study vertebrates and these researchers have suggested that the females can engage in a number of mating (polyandry). This is done in order to 'hedge' their bet that their initial mate or mates will provide a sufficient number of high - quality gametes (Double and Cockburn, 2000; Fox and Rauter, 2003; Whittingham et al., 2010).

In fruits of *Mimulus ringens*, the outcross seeds are sired by one to eight pollen donors (Mitchell et al., 2005). There has been demonstration given by Karron et al. (2006) that the sequential pollinator probes to individual flowers tend to enhance the mate diversity within fruits significantly.

If the timing of stigmatic receptivity is regulated by the plants, then the stigma closure or floral longevity, they could have the potential to influence mate diversity within fruits (Bernasconi, 2003; Lankinen and Madjidian, 2011).

It is important to have more research which is required in order to understand the way in which the variation of mate diversity takes place within and among populations, the factors which regulate the mate diversity and the fitness consequences of variation in mate diversity in natural populations.

5.7. THE FUTURE OF PLANT MATING SYSTEM STUDIES IN THE ERA OF NEXT-GENERATION SEQUENCING

There has been research a generation ago on plant mating systems which further highlighted and further blossomed with the widespread availability of co - dominant genetic markers (Brown and Allard, 1970).

This is done while facilitating studies of outcrossing rates as well as patterns of paternity. In today's scenario, the plant evolutionary biologists are more adapting to some other revolution in molecular technology, NGS that is next-generation sequencing. This again opens up fresh areas of inquiry.

By considering the advantage of some of the current developments in genome sequencing technology, now, the humans have the ability to address these and some other previously intractable questions.

The researchers are increasingly able to genotype hundreds of markers within populations of non-model organisms along with next-generation sequencing technologies, and also, they tend to probe the genetic architecture of population as well as individual traits critical for answering important questions in relation with the plant mating systems.

The benefits of NGS are being considered across a wide range of fields which include genomics (Hawkins et al., 2010), human disease (Day-Williams and Zeggini, 2011), genetics (Metzker, 2010) and crop genetics and breeding (Varshney et al., 2009).

There is much excitement related to the potential for NGS in order to benefit fields closely which are allied to plant mating systems, which include the genomic ecology as well as evolutionary biology (Hudson, 2008), conservation genetics (Allendorf et al., 2010; Ouborg et al., 2010) and molecular ecology (Ekblom and Galindo, 2011).

5.7.1. Rapid Cost-Effective Marker Development for Mating System and Paternity Analysis

There is one far-reaching benefit of NGS and that is already being realized across a number of fields which is rapid, cost-effective as well as genome-wide discovery of variable genetic markers like the microsatellites (Dalca and Brudno, 2010; Davey et al., 2011). Gardner et al. (2011) furthermore offer a helpful procedural guide for the ecologists planning in order to use NGS for marker development.

Moreover, as the technology improves, the simultaneous genotyping of large panels of microsatellites across a number of individuals will be possible in any of the species for a fraction of current costs. Therefore, for the first time, the NGS will enable cost - effective paternity analysis in large natural populations.

This is such a capability which could even enable determination of pollen donor representation in the pollen load on a single stigma. There can be comparison to realized paternity for whole seed families which could result in exciting insights into pollen competition as well as reproductive incompatibilities.

5.7.2. Contributions of NGS for Phylogenetic Analysis

Onto a robust phylogeny, the mapping of mating system traits can further reveal important evolutionary insights. Let's take an example. Ferrero et al. (2012) discovered phylogenetic evidence which is for the independent evolution of heterostyly as well as incompatibility in the Boraginaceae, further contradicting the expectation of strong connection in between these traits.

Unfortunately, most of the time, a lack of phylo-genetically informative markers tends to prevent the construction of robust phylogenies which is said to be the first step in this process. Particularly, this is true in recent time evolved groups which can also offer the most potential in order to explore the dynamic evolutionary processes (Hodges and Derieg, 2009; Peakall et al., 2010).

Moreover, the NGS can assist phylogenetic analysis in at least two ways. First and foremost, rapid as well as low-cost phylogenetically informative genetic markers can be found through the transcriptome sequencing which tends to target the coding genes (Ekblom and Galindo, 2011).

The genome-wide phylogenetic analysis can be achieved by using RAD-seq that is restriction-site associated DNA sequencing. This in turn offers the potential in order to detect the 1000s of polymorphisms at the same time across a number of individuals in a single NGS run (Davey and Blaxter, 2011; Davey et al., 2011).

5.7.3. Candidate Gene Discovery

Beyond the time of marker development *per se*, the NGS tends to offer exciting and new opportunities in order to track the candidate gene discovery fast (Bräutigam and Gowik, 2010). Whereas, on the other hand, the genetic bases of SI that is self-incompatibility mechanisms and the loss of SI are increasingly well understood in the model systems like *Arabidopsis* (Shimizu et al., 2011). The NGS can open the door for rapid progress in candidate SI gene discovery in non - model systems.

In the similar manner, human tend to anticipate NGS which will further facilitate progress on the genetic basis of sex determination in plants. This is central to the understanding of evolution of separate sexes. In addition to this, in mating system traits, identifying the loci underlying variation is becoming increasingly accessible outside model systems which are done through developments in NGS.

The restriction-site associated DNA sequencing techniques can further generate a wealth of data across the whole genome for a sample of individuals in a population. Within the population sample, when individual sequences are grouped by some measured phenotype, loci in connection with the measured trait which can be identified (Davey and Blaxter, 2011).

The theory for inferences which are marker - based of the heritability of quantitative traits is already well developed (Ritland, 1996) and, in conjunction with the developing measures of relatedness for population data which gleaned from massive batteries of markers (Ritland, 2011; Browning and Browning, 2010), the underlying genetic architecture of floral traits is now within reach.

5.7.4. Rapid Chloroplast Genome Sequencing

Along with next-generation sequencing capability, the whole chloroplast genome is becoming an increasingly accessible tool in order to be used in both population as well as community studies. Let's take an example. Doorduin et al. (2011) further reported the complete chloroplast genome sequence of 17 individuals within a single pest plant species, *Jacobaea vulgaris*.

Despite the fact that only recently recognized for its potential so as to reveal genetic variation within species (Ebert and Peakall, 2009), when it is combined with predominantly uni-parental transmission, the genetic variation at the chloroplast is ideally suited in order to tease apart pollen and seed-mediated gene flow.

There have been studies which further use the individual levels of cpDNA variation which will also highlight the patterns of both maternity as well as paternity in dispersed seeds and juvenile plants in natural populations.

Furthermore, there have been suggestions provided by Metzker (2010) that the potential applications of NGS are primarily limited by 'one's imagination', and also, there can be little doubt that NGS tends to offer powerful as well as exciting new opportunities in order to advance the field of plant mating systems.

However, neither the NGS is a panacea, nor is it a substitute in order to employ simple, yet elegant experimental designs, or a justification so as to abandon the strategically chosen 'model' systems.

Ultimately, notwithstanding the promise of NGS, there remain a number of outstanding questions in relation with the plant mating systems that can further be investigated effectively without the need to conduct either genetic or molecular analysis at all.

5.8. CONCLUSION

It is considered that the prevalent and substantial are among the population variation which is in the outcrossing rates highlighting all the need for cation which in both the estimation and interpretation of the outcrossing rates in the flowering plants.

Estimation from the single population should not be used for characterizing the mating systems of an entire species. It has been noticed from last several decades that the research on the causes and consequences of the plant mating data, variations, models, and hypotheses that have been proliferated.

There is an availability of unifying framework that is used for explaining the frequency that is of mixed mating which remain elusive. Further, the insights in the future work will often require the additional multi-population with the relevant ecological data.

Therefore, in order to move them beyond the correlative explanations. Therefore, a careful experimental study that basically combines the ecological manipulations and sampling are over the multiple seasons and is considered to be needed for distinguishing the relative influence of the ecology and genetics on the evolutionary enigma of the mixed mating.

REFERENCES

1. Bicknell, R., 2004. Understanding Apomixis: Recent Advances and Remaining Conundrums. *THE PLANT CELL ONLINE*, [online] 16(suppl_1), pp. S228-S245. Available at: <http://www.plantcell.org/content/16/suppl_1/S228> [Accessed 14 June 2020].
2. Encyclopedia Britannica. n.d. *Cross-Fertilization | Biology*. [online] Available at: <https://www.britannica.com/science/cross-fertilization> [Accessed 14 June 2020].
3. Encyclopedia Britannica. n.d. *Self-Fertilization*. [online] Available at: <https://www.britannica.com/science/self-fertilization> [Accessed 14 June 2020].
4. Karron, J., Ivey, C., Mitchell, R., Whitehead, M., Peakall, R. and Case, A., 2011. New perspectives on the evolution of plant mating systems. *Annals of Botany*, [online] 109(3), pp.493-503. Available at: <https://www.ncbi.nlm.nih.gov/pmc/articles/PMC3278297/> [Accessed 14 June 2020].
5. Karron, J., Ivey, C., Mitchell, R., Whitehead, M., Peakall, R. and Case, A., 2011. New perspectives on the evolution of plant mating systems. *Annals of Botany*, [online] 109(3), pp.493-503. Available at: <https://academic.oup.com/aob/article/109/3/493/130434#83958912> [Accessed 14 June 2020].
6. Pierre-Olivier, C., n.d. *The Evolution of Plant Mating System: Is It Time for A Synthesis?* [ebook] Available at: <https://pdfs.semanticscholar.org/138c/b190906444af48718fdc2ed97fca9171ac95.pdf%20%20%20%20%20-1> [Accessed 14 June 2020].
7. Xie, N., Chen, L., Dong, Y. and Yang, H., 2019. Mixed mating system and variable mating patterns in tropical woody bamboos. *BMC Plant Biology*, [online] 19(1). Available at: <https://bmcplantbiol.biomedcentral.com/articles/10.1186/s12870-019-2024-3> [Accessed 14 June 2020].
8. Allendorf et al., 2010
9. Ashman T-L, et al. 2004
10. Ashman, 2000
11. Asker and Jerling, 1992
12. Barrett S, Harder L, 1996
13. Barrett, 2008

14. Bell et al., 2005
15. Bernasconi et al., 2004
16. Bernasconi, 2003
17. Bicknell et al., 2003
18. Bräutigam and Gowik, 2010
19. Brown and Allard, 1970
20. Brown, 1990
21. Brunet and Sweet, 2006
22. Carman, 1997
23. Case and Ashman, 2009
24. Czapik, 1994
25. Dalca and Brudno, 2010
26. Darwin, 1877
27. Davey and Blaxter, 2011
28. Davey et al., 2011
29. Day-Williams and Zeggini, 2011
30. Dickinson, 1998
31. Doorduin et al., 2011
32. Double and Cockburn, 2000
33. Ebert and Peakall, 2009
34. Ekblom and Galindo, 2011
35. Falconer, 1981
36. Ferrero et al. (2012
37. Fisher (1941
38. Fox and Rauter, 2003
39. Gadella, 1991
40. Gardner et al. (2011
41. Griffin and Eckert, 2003
42. Grimanelli et al., 2001
43. Harder and Barrett, 1996
44. Hawkins et al., 2010
45. Heenan et al., 2002
46. Herlihy and Eckert, 2004

47. Hodges and Derieg, 2009
48. Holsinger K, 1991
49. Horandl, 1998
50. Hudson, 2008
51. Karron and Marshall, 1993
52. Karron and Mitchell, 2012
53. Karron et al., 2009
54. Kelly and Willis, 2002
55. Knight T, et al. 2005
56. Krahulcovà et al., 1999
57. Lande R, Schemske D, 1985
58. Lankinen and Madjidian, 2011
59. Leclerc-Potvin and Ritland, 1994
60. Metzker, 2010
61. Mitchell et al., 2009
62. Moeller, 2006
63. Mogie, 1992
64. Nason and Ellstrand, 1995
65. Nogler, 1984
66. O'Connell and Eckert, 1999
67. Ouborg et al., 2010
68. Peakall et al., 2010
69. Philipson, 1978
70. Plitman (2002
71. Richards, 1997
72. Ritland, 1996
73. Ritland, 2002
74. Schemske D, Lande R, 1985
75. Shaw et al., 1981
76. Shimizu et al., 2011
77. Shuster S, Wade M, 2003
78. Smith, 1841
79. Uyenoyama et al. 1993

80. Uyenoyama, 1986
81. Varshney et al., 2009
82. Vogler D, Kalisz S, 2001
83. Waller, 1993
84. Waser, 1993
85. Whittingham and Dunn, 2006
86. Whittingham et al., 2010
87. Winkler (1908
88. Wright (1922) and Fisher (1941)
89. Yahara, 1992
90. Yasui, 1998
91. Young et al., 2012

6

Principles and Practice in Breeding Perennial Horticulture Crops

CONTENTS

The new perennial species of crop development is supposed to be reviving in the form of a favorable approach for changing the elementary nature of ecosystem agricultural processes. The arguments pertaining to the ecology for the purpose of growing perennial crops in polycultures would be vigorous. However, in the recent times, perennial herbaceous grain crops would not be found in the agricultural landscape.

It is not that perennial herbaceous species would be nonexistent in nature, uncountable perennial grasses, legumes, and other plants with broad leaf would be found. Rather, for a number of reasons, farmers from the early times concentrated on cultivation and domestication of annuals and perennial herbs are considered to be mostly overlooked. At present there would be immense scope for exploring a different agricultural path.

Constructed on modern knowledge related to plant biology and genetics which was supposed to be lacking in early farmers, with the usage of a rapidly increasing skillset inclusive of sophisticated genomic and analytical methods, development of perennial grain crops would be possible. For assembling varied agroecosystems such crops could be used, for soil regeneration and capturing other vital functions of the ecosystem.

As the name suggests, the chapter focuses on the various methods and ways of breeding perennial crops. It further includes breeding different perennial grains providing an overview on their chromosomal build up and strategies for improvising the yield. Direct domestication is another point that has been elaborated in the chapter along with a reference to few of the crops associated with it.

Additionally, the chapter highlights on the fruit crops specifying few of the Mediterranean, Central Asian and Chinese and Southeast Asian fruits. Lastly, few technical glitches related to fruit breeding have also been included in the chapter.

6.1. INTRODUCTION

It is considered that humans would have started the consuming seeds which they obtained from several wild grasses for more than 23,000 years BP (Weiss et al., 2004) and there would be potential evidences supporting the same. Moreover, there would be added evidences that prove the consumption of cereals by humans that would have started for more than 100,000 years ago (Mercader et al., 2009). The domestication of annual species progressively transformed the humans and cereals relationship from optional to necessitate, at a minimum of six originating centers across the globe 5-10,000 years BP (Gepts, 2004). The annual grass species would be comprising of wheat, rice, barley, maize, millet, and sorghum.

Additionally, apart from grass seeds, pulses, or seed-crop legumes, like beans or lentils, are supposed to have been domesticated as well in the same major hubs of crop domestication (Gepts, 2004, Harlan, 1992). Hence, the synergy among the humans and annual cereal, pulse, and oilseed crops, or in other words , grains turned out to be outstandingly effective with the onset of domestication.

Several geographers, anthropologists, archeologists, geneticists, and agronomists tried constantly to research on the question as to why there would have been a gradual change from a hunted and foraged food diet to an extremely yearly cultivated grain based diet (Gepts, 2004, Cohen, 2009, Harlan, 1995).

Multiple factors would have been usually associated with the theories, like scarcity of food, changing climate, shortfall of resources, grain storage and socio-economic structural change with a much inactive lifestyle. Yet, significantly, ecological nurturing for long-term would be from the record of probable causes justifying human commitment towards farming of yearly grains.

Certainly, the natural ecosystem conversion would be characterized through varied periodic vegetation to agroecosystems that would be dependent on the regular removal of all vegetation for the establishment of annual grain species.

It is supposed to be dedicated towards a strategy of food procurement that would be tricky. There would be ample examples in the history of annual production of grains that would have led to huge degradation in soil developing into a major cause for social breakdown. (Carter et al., 1955, Hillel, 1992, Montgomery, 2007, Gomiero, 2016).

Figure: 6.1: Agriculture in Ancient Times.

Source: Image by Wikipedia

Besides, despite investing a lot and extensive campaigning to stop erosion and degradation of the soil throughout the last century, a 2015 FAO report with more than 200 soil scientists from across the globe have concluded: though few locations would provide optimistic results, however, the profound result drawn from regional assessment would be that most of the world's soil resources are considered to be of only good, poor, or of very poor state.

At the global scale, the biggest risk to soil functioning would be soil erosion, losing organic carbon from the soil as well as nutrient imbalance. Without collaborated actions from individuals, private sector, governments and international organizations, the existing perspective for such situation would deteriorate (Food and Agriculture Organization of the United Nations, 2015).

An enduring risk towards the existence of the species would be posed through soil erosion that would lead to loss of soil resource which would be caused as a result of trying to grow abundant annual grains to suffice the demands of human diet.

Nevertheless, in ecoservice disservice, erosion solely would not be the reason leading to perennial ecosystem of high diversity to an annual agroecosystem of low diversity (Crews et al., 2018). For example, nitrification inhibitors, variable rage fertilizer application as well as patented, herbicides tolerant crops that are genetically engineered.

> For agricultural improvement, several input and capital-concentrated approaches are supposed to have been developed.

Though the above mentioned as well as other solutions would be usually successful in reducing distinct ecosystem disservices, yet it would further confine the farmers to a technological treadmill steering the money out of their farms to the producers and marketers of the equipment and inputs.

There would a valuable cause for expecting several ecosystem services being carried out through the natural ecosystems predating agriculture for improvement, by means of crop species perennializing and expanding the diversity of crop at the field, farm, and regional level.

6.2. PERENNIAL CROPS

The more permanent crops that would necessitate many growth cycles prior to producing fruit, would be generally known as the perennial crops. Their need for the cold winter temperatures before the following spring for flowering, is considered as one of the characteristics of such plants.

Moreover, there would be an evidence of few perennial crops not getting the required chilling hours to their optimum, resulting in limited flowering and eventually fruit set. It would be due to the warm temperatures during winter (due to climatic changes when winters are not as cold as it is supposed to be).

Moreover, promotion of early budding and developing flower as a result of warm winters and damaging those flowers, fruits, or nut formation by the events of frost occurring in the late season, would be an added concern. The value of perennial crops would be due to the consolidation of their entire production and the harvested product quality. For example, size of an apple or peach, the red blush on an apple, the amount of oil in olives or a specific vineyard producing a distinct quality of wine.

Naturally, the production of perennial crop would be for a longer period of time with significant production issues due to climate change. Once established, the production area location being less flexible and the production cycle being calculated in decades instead of a particular season of growth, would be considered as the reason behind it.

An excellent example of showing the effect of climate stress on the growth all over the year and in the course of years would be of an apple. Floral initiation (occurs in June – July) is supposed to be affected with the temperatures going high at the time of the year prior harvesting and would be reducing the quantity and potency of the floral buds.

During dormancy the tree would be exposed to severe cold temperatures, destroying the buds and the erratic periods of high temperatures would result the flower buds becoming more vulnerable towards winter damage before the recommencement of growth in spring.

There could be a possibility of flowers emerging in the spring, however, with the occurrence of frost, it is supposed to be killed. High temperature exposure would lead to lower cell division producing smaller size of fruit, at the time of its development in the spring.

In the months of summer, with temperatures going high might result in damages occurring due to sunburn and low harvest quality, fast maturity, reducing the firmness and color development of the fruit and making the fruit less suitable for storing it for short or long-term.

Figure: 6.2: Frost Damages Perennial Flowers.

Source: Image by Pxfuel

A water scarcity during the same period, would result in aggravating the effects. Occurrence of climate stress numerous times at the time of growing season, would usually not be recognized, yet it is considered to be a common event in perennial crops.

6.3. FRUIT CROPS

6.3.1. Mediterranean Fruits

1. Date Palm

the date palm (*Phoenix dactilifera*, Arecaceae) would be a dioecious palm, believed to be originating from Northern Africa across the Arabian Peninsula tonorthern India (Goor and Nurock 1968; Zohary and Spiegel-Roy, 1975; Smartt and Simmonds 1995).

For the fresh or dried consumption of its sweet fruit, its valuable leaves and wood andfor its longevity, the date palm was supposed to be valued in the ancient times, from India to North Africa.

Figure: 6.3: Date Palm Farm.

Source: Image by Flickr

The date palm is considered to have been transformed into a sacred tree, due to its several virtues, which would be referred symbolically in Babylonian and Assyrian iconography. It might have been considered as the first cultivated fruit being deep-rooted in the Middle East throughout the Bronze Age.

A lot of speculations would be there to decide on the exact origin of date cultivation. According to Vavilov, the northeastern African mountains in Ethiopia and Eritrea would be considered to be the place of origin of date. However, evidence of the cultivation occurring in the Lower Mesopotamian Basin for the first time, would be available.

As per archeological proofs the date stones would be place in Ubaidian horizon (about 400 BCE) in Eridu, lower Mesopotamia. From the time of pollination to harvest, the date palm would be accustomed to long, relatively high summer temperatures, with low rainfall as well as less humidity. However, it is an underground moisture source.

Every fruit in conditions of dry desert, a several wild species not producing suckers and might being accustomed to rainy conditions. Irrigation would be necessary as the plant being a monocot and lacking in a deep root system.

As per an Arab proverb, the date palm would be described as "its feet in running water and its head in the fire of the sky."

Unlike its wild relative, the seed of a date palm would be able to easily propagate a cultivated date palm as well as at the base of the plant could be propagated vegetatively, from offshoots (suckers). One of the main characteristics of domesticates would be the clear production of offshoots.

Olive

Being an evergreen treeolive would be adapted to the Mediterranean basin climate with the characteristics of slow growth and longevity which is supposed to be an important feature of the particular climate (Smartt and Simmonds 1995). With the archeological proofs going back to 12,000 years ago, it could be said that the indigenous varieties would be still extensively found (Blazquez, 1996).

Figure: 6.4: Olive Plantation.

Source: Image by Pxfuel

Around 6000 years ago the cultivated olive originated in Asia Minor, however, was usually the Babylonians and Assyrians for whom sesame and walnut was the source of oil, were unaware of it. For a long time, olive was known in Syria and the Holy Land and was popularized in Egypt from Syria between 3000, 1800 BCE (Goor and Nurock, 1968).

The ancient Elba city in northern Syria, by 2500 BCE, had fields with 500 and 1000 olive trees that produced and exported different varieties of olive oil. Hittite texts would have highlighted the cultivation of olives in Anatolia as well as sent to Egypt in the form of an import fruit since the Ramses time (1197-1165 BCE).

Where the olive oil would have been used as an illuminating agent and a skin emollient to treat cracks and sunburn. The olive is supposed to have reached Italy in the 7th to 6th century BCE, after travelling from Egypt to Carthage in North Africa.

Grape

Wild grapes belonging to the Old World (Vitis sylvestris, Vitaceae) would be considered aboriginal to the South Caspian belt, Turkey, and the

Balkans. They were supposed to be distributed extensively in the northern Mediterranean area with the Black and Caspian Seas being included (Zohary and Spiegel Roy 1975; Smartt 16 and Simmonds 1995; Reisch and Pratt 1996). In contrast to olives, wild grapes would be dioecious, perennial, forest vines thriving in cooler and excessive humid conditions.

Figure: 6.5: Wild Grapes.

Source: Image by Flickr

Harvesting wild grapes had been preceded by domestication for long and would have been proved by carbonized seeds in several prehistoric sites in Europe. In Near East, existence of wine grape would be dated as early as the 8th millennia BCE, but due to deficiency of winter hardiness and low adaptation to high summer rainfall, its eastward expansion had been limited.

V. vinifera, or the wild grape, towards 5000 BCE and probably before, would have migrated from Anatolia to Syria and to the Holy Land thereafter. In Mesopotamia, the Holy Land, Syria, Egypt and the Aegean, the Bronze Age domestication are supposed to have been found. There would be proof of vessels for storing wine as well as of raisins by the 2nd millennium BCE.

Fig

A gynodioecious species with monoecious wild varieties (caprifig) and pistillate domesticates, commonly known as the fig (*Ficus carica*, Moraceae) (Condit 1947). In the Mediterranean basin, figs would beregarded one among the classical fruits and borne on small trees. At different Neolithic and late Neolithic sites (Zoharyand Spiegel-Roy, 1975) and from Bronze Age sites during the 4th millennium BCE, gives an idea of fig cultivation.

With olive and grape, domestication is considered to be typically contemporary in the Eastern Mediterranean basin; King Urukagina ca 2900 BCE mentioned it making the Assyrians aware of it as early as 2000 BCE. In the 2nd and 3rd millennia BCE fig cultivation in Canaan had ben indicated through Egyptian records (Goor and Nurock 1968) along with several ancient paintings of fig. Propagation of fig would be simply done by seeds as well as by cutting or layers. In the 3rd century BCE, Cato described propagation through grafting and budding.

Pomegranate

Pomegranate (*Punicagranatum*, Punicaceae) being cultivated for 5000 years, a Bronze age fruit and an aboriginal to the southern Caspian belt (Iran) and northeast Turkey (Goor and Nurock 1968; Zohary and Spiegel-Roy 1975). Domestication of the fruit would be indicated through its presence in Syria, the Holy Land, Egypt, and Mesopotamia.

Figure: 6.6: Pomegranate Plantation.

Source: Image by Openphoto.net

Going by its Latin name Punica that would be referring to Carthage, at one time the fruit was known as the apple of Carthage. Nimrud, Arad, and Jericho would have its remains. Almost 1600 BCE before the 19th Dynasty, at the time of the Semitic people inflow (Hyksos), the pomegranate had been introduced.

Typical illustration of the fruit would be there in Egyptian art from the New Kingdom and largely found in Jewish and Greek art. By the cuttings of hardwood and softwood, the pomegranate would be easily propagated.

Increased size of the fruit and sweetness and shifting towards clonal propagation would be accompanying domestication. The ones with degenerated embryos would be known as seedless.

6.3.2. Central Asian Fruits

1. Pome Fruits

Pome fruits belonging to the Rosaceae consisting of apple, pear, quince and medlar, among which the apple would the most significant tree fruit belonging to the temperate world. To disrupt dormancy for adequate growth, an annual period of chilling would be necessary for the trees.

Prior to germination, a period of cold stratification would be required by the seeds, and necessarily be detached from the placental tissue of the fruit for germinating.

2. Apple

Cross-pollinating and major cultivars being self-incompatible would be the characteristics of apple. The apple and pear trees maximum cuttings would not be rooting from cuttings to an extent vegetative propagation would not be easy without grafting. There would be a distribution of 24 key Malus species in Europe, Central Asia, and Eastern Asia and three in North America (Way et al. 1990).

Majority of the wild apples are supposed to be of a smaller size and bitter in taste. However, the domestic apples that would be large, sweet-smelling (Malus ×domestic) would certainly be originating in Central Asia, particularly in Almaty, Kazakhstan (Dzhangaliev, 2003) and in the ancient times it was introduced in the west via Persia.

Figure: 6.7: Apple Orchard.

Source: Image by Pikist

3. Pear

Apart from following two different paths of domestication, the story of the pear would be similar to that of the apple. The European pears were produced by *Pyrus communis*, native to Central Asia. Intense flavor, innumerable shapes and sizes and softening of the flesh after the climacteric would be few of its characteristics.

Homer described the European pear among one of the fruits in the garden Alcinöus, while the Roman agricultural writers like Cato, Varro and primarily, Pliny would be describing several varieties (Hedrick, 1921).

Figure: 6.8: Pear Harvest.

Source: Image by Pxfuel

The crisp, sweet fleshed *Pyrus pyrifolia*, *P. bretschneiderii* and *P. ussuriensis* fruit would be round, fragrant with crisp flesh, in eastern Asia, which at present would be called as either Asian pear or would have a Japanese name nashi (Watkins 1995; Bell et al. 1996). For 4000 years the pear is supposed to have been known in China (Wang, 1990).

4. Stone Fruits

For almond, apricot, cherry (sweet and tart), peach (and nectarine) and plum the genus Prunus, Rosaceae would be the source. All others would be soft-fleshed temperate fruits popular for their delicious flavors, except almond, would be cultivated for the purpose of its seed.

With Eastern Asia, Europe and North America being the secondary centers, Prunus is supposed to have been originating in Central Asia (Watkins, 1995). Apricot, cherry, and plum being introduced into the Mediterranean basin as well as its association with Alexander the Great (356-323 BCE) invasion wherein, he conquered Persia subsequently moving towards east through Turkistan, Afghanistan, Pakistan, and north west India till river Indus.

5. Almond

Starting from Turkey and Syria into Caucasus and into the Tian-Shan deserts as well as the Hindu Kush mountains, the almonds would be growing wild over southwest and Central Asia (Zohary and Spiegel-Roy 1975).

On the basis of biblical literature, almonds might have been introduced during the 2000 BCE with few almonds remains considered to be found in the 18th dynasty in Egypt (1550 BCE). Until the time of Romans, it would not look like being extensively cultivated in Egypt. Around 200 to 300 BCE it was supposed to have been introduced into Greece, slowly reaching the whole Mediterranean basin.

Figure: 6.9: Almonds.

Source: Image by Flickr

6. Apricot

The apricot (*Prunus armeniaca*) history would be dated 5000 years back in China, with Emperor Yu (2205-2198 BCE) being first attributed, along with other references in 658 BCE. In 406-250 BCE higher quality orchards are supposed to have been described (Faust et al., 1998). Around 600 CE apricots grafting would have begun and distinct cultivars being developed after that time.

In China, there would be minimum 11 cities with xing (apricot) included in their name. The closely associated *P. mume*, or the Japanese apricot significant for blooming early exclusively significant for their gnarled branches, abundant early flowering, and fragrance. Paintings related to such species would be a distinct form of art.

6.3.3. Chinese and Southeast Asian Fruits

1. Peach

Domesticated prior to 3300-2500 BCE, peach (*Prunus persica*, Rosaceae) is considered to have been originated in China (Faust, 1995). Peach culture would be going back to 2000 BCE years and presently China would have numerous cultivars (Wang, 1985). The Chinese literature dated before 1000

BCE would have references of peach which would even be entwining in Chines mythology and folklore.

In various parts of China several forms were supposed to have been selected. The peach tree would be characterized by extreme chilling requirements, long internodes, single flower buds, vertical branching, large flat leaves and mostly cling or semi-cling fruit.

2. Citrus

Rutaceae have six genera among which Citrus, Poncirus (*Trifoliate orange*) and Fortunella (Kumquat) include commercial benefit. A varied group of evergreen Old World fruits would be included in the genus Citrus and southeast Asia (eastern India, Burma, and southern China) being the place of origin, presently grown worldwide over the tropics and subtropics (Soost and Roost, 1996).

Citron (*C. medica*), lemon (*C. limon*), lime (*C. aurantifolia*), pummelo (*C. grandis*), sour orange (*C. aurantium*), sweet orange (*C. sinensis*), mandarin (*C. reticulata*) and grapefruit (*C. paradisi*) would be included in the chief horticultural groups. Apart from grapefruit, for all others domestication would have occurred prior 1000 BCE.

3. Banana and Plantain

Bananas and plantains (Musa species, Musaceae) would be aboriginal to southeast Asia and the Pacific (Simmonds, 1995). They would be huge, herbaceous, perennial plants that would have a pseudo stem with firmly whirled sheaths of leaf whereby inflorescence would be emerging. The fresh fruit varieties would be named bananas, whereas, the cooking varieties would be referred as plantains.

Corms (the true stem) with their offshoots would be propagating them vegetatively. Though, the initial evolution and migration would not be known, however, it would be ancient. Banana would be denoted as Theophrastus and would have undoubtedly originated thousands of years ago.

Prior to European contact, west Africa would have been the destination of bananas, and from there a handful of clones is supposed to have reached the New world in the later part of 15th century. By the end of 16th century they would have been sweeping the tropics.

4. Mango

Mango (Mangiferaindica) and pistachio (*Pistacia vera*) native to East Asia, would be regarded as important crops of the Anarcardiaceae, while the cashew (*Anacardium accidentale*) aboriginal to tropical Brazil. Mango being the major popular fruit in India, would be ranking fourth most important fruit crop globally.

From nearly 90 countries more than 25.1 million tonnes of mango are considered to have been produced in 2001, with 77% of the produce being from Asia (GalánSaúco, 2003).

Figure: 6.10: Mango Harvest.

Source: Image by Pxfuel

The Indo-Burma region would be the place of origin of mango that would be growing wild in the northeast Indian forestand is considered to be an allopolyploid between *M. indica*an *M. sylvatica*. Famous for 4 to 6 thousand years, the mango travelled the Indian subcontinent reaching the Malaya in around 5th century BCE (Purseglove, 1968), in 10th century East Africa and in the beginning of the 19th century the New World.

In 1861 it had become successfully known in Florida. Mango would be eaten in the fresh, dried, or processed form and would be commonly used in chutneys. For mango lovers, it is supposed to be a fruit that the peach would always desire of becoming.

5. Kiwifruit

An example of an ancient species of fruit that is considered to have been domesticated in the 20th century would be the kiwifruit (*Actinidia deliciosa*, Actinidiaceae, 2n=174, x=29) being regarded as a dioecious vine. It is considered to have been emerged from mihautao (monkey peach) which would be a gathered fruit and had been liked in China for a long time. However, it's been collected and not cultivated (Ferguson, 1990).

It would be known as the Chinese gooseberry in the United States, after being introduced to the United Nations and New Zealand by E. H. (Chinese) Wilson in the beginning of the 20th century. Even though it maintained inquisitiveness in the United States, the growers and nurserymen in New Zealand have successfully domesticated the crop by adequately selecting male and female clones, along with developing methods for cultivation.

6.4. TECHNICAL PROBLEMS IN FRUIT BREEDING

Majority of woody perennial fruit plants would be outbreeders and extremely heterozygous while not breeding the original seed form. Consequently, the subtle gene combinations consisting a decent cultivar would necessarily be maintained by vegetative propagation. Inherited polygenic and some simple Mendelian characters would be the most common traits for making a good fruit.

Woody perennial plant seedlings would consist of a prolonged juvenile phase when formation of flowers would generally occur. The minimum generation times would be ranging from 4-15 years, being conditional to the species, with modern methods. Additionally, for evaluating orchards or vineyards of advanced selections, the time required would be of another 5 years. Self-incompatibility, polyploid and parthenocarpy would be other biological restrictions in fruit breeding.

Lastly, after raising a progeny there would be issue of selection criteria that would be encountered by the fruit breeder. Generally, quality would be the main criteria for breeding fruit crops and not yield. For that there would be certain easy objective tests of fruit quality. Finally, prevalence of certain orthodox forces would typically try to restrict the introduction of new cultivars regardless of quality.

Some cultivars are supposed to be strongly established in the market-place (e.g. Granny Smith) and consumers would not be much interested in fruits with which they are not accustomed to. New cultivars would not be generally appreciated by wholesalers as it would make marketing more complex. Introducing new genotypes in certain crops, preferably grapes, would be obstructed by schemes, like "Appellation d'Origine Controlee" designed for the purpose of stabilizing and quality and strengthening the image of existing products.

6.5. CONCLUSION

Crop plants would be considered to be the most valuable asset from the pre historic period (Harlan 1992; Diamond 2002). The means, place, and the time of occurrence in relation to domestication of crop plants would be gradually revealing if not being completely comprehended (Camp et al. 1957; Smartt and Simmonds 1995; Gepts 2003).

In few cases, the origins of wild and domestic plants would be vague due to their extreme genetic distance, like maize. The origins regarding early grains (wheat, maize, rice) and pulses (sesame and lentil) which would have been domesticated during the Neolithic times would have been a great interesting topic for contemplation which would be solved with new proof based on molecular biology (Gepts, 2003).

Between 6000 and 3000 BCE, in the later part of Neolithic and Bronze Ages, domestication of ancient Mediterranean fruits (date, olive, grape, fig, pomegranate) occurred (Zohary and Spiegel-Roy 1975).

Domestication of fruits, like citrus, banana, pom fruits (apple, pear) and stone fruits 9almond, apricot, peach, plum) occurred in Central and East Asia and was introduced in the West, during the ancient times. In the 19th and 20th centuries some fruits were domesticated (kiwifruit).

REFERENCE

1. 2016. Breeding of Field & Horticultural Crops. [ebook] p.3. Available at: <https://www.agrimoon.com/wp-content/uploads/Breeding-of-Field-Horticultural-Crops.pdf> [Accessed 11 June 2020].
2. Janick, J., 2005. The Origins of Fruits, Fruit Growing, And Fruit Breeding. [ebook] p.5. Available at: <https://hort.purdue.edu/newcrop/origins%20of%20fruits.pdf> [Accessed 11 June 2020].
3. Mullins, M., 2006. The Regional Institute - Publications. [online] Regional.org.au. Available at: <http://www.regional.org.au/au//asa/1980/invited/genetic-exploitation/mullins.htm> [Accessed 11 June 2020].
4. North, C., 1979. Plant Breeding and Genetics in Horticulture. [ebook] p.2. Available at: <https://link.springer.com/content/pdf/bfm%3A978-1-349-03707-0%2F1.pdf> [Accessed 11 June 2020].
5. Purseglove, J., 1970. Outlines of Perennial Crop Breeding in the Tropics. F. P. Ferwerda and F. Wit, Eds. Veenman, Wageningen, The Netherlands, 1969. xvi, 512 pp., illus. Paper, N.fl. 49.50. Landbouwhogeschool Miscellaneous Papers, No. 4 (1969). Science, [online] 170(3959), p.2. Available at: <https://edepot.wur.nl/455436>.
6. Sciencedirect.com. 2020. Perennial Crop - An Overview | Sciencedirect Topics. [online] Available at: <https://www.sciencedirect.com/topics/agricultural-and-biological-sciences/perennial-crop> [Accessed 11 June 2020].
7. Anamthawat-Jonsson et al., 1997
8. Anamthawat-Jonsson et al., 1999
9. Anamthawat-Jonsson, 1996
10. Anaya, 1999
11. Anderson, 2018
12. Banks et al., 1992
13. Banks et al., 1993
14. Bar-Yosef, 2004
15. Becker et al., 1991
16. Bell et al. 1996
17. Bell, 2018
18. Bender, 2002

19. Bennett et al., 1998
20. Blazquez, 1996
21. Bodrov, 1960
22. Brummer, 2002
23. Cai et al., 1998
24. Cai et al., 2001
25. Camp et al. 1957
26. Carter, 1955
27. Carton, 2018
28. Cattani, 2018
29. Cauderon, 1979
30. Cohen, 2009
31. Condit, 1947
32. Cox, 2002
33. Cox, 2010
34. Cox, 2018
35. Crews, 2018
36. Dale, 1955
37. DeHaan, 2010
38. Derzhavin, 1960a
39. Devos et al., 1993
40. Dewey, 1984
41. Diamond, 2002
42. Dorffling, 1988
43. Dvorak, 1970
44. Dvorak, 1976
45. Dzhangaliev, 2003
46. Faust et al., 1998
47. Faust, 1995
48. Ferguson, 1990
49. Food and Agriculture Organization of the United Nations, 2015
50. Frels, 2018

51. GalánSaúco, 2003
52. Gazza, 2018
53. Gepts 2003
54. Gepts, 2004
55. Gomiero, 2016
56. Goor, 1968
57. Gordon-Werner 1988
58. Griffin, 1981
59. Gustafson et al., 1989
60. Harlan, 1992
61. Harlan, 1995
62. Hayes, 2018
63. Hedrick, 1921
64. Hillel, 1992
65. Hodosne-Kotvics et al., 1996
66. Holland, 2002
67. Jackson and Jackson, 1999
68. Jackson, 2002
69. Jauhar, 1992
70. Jauhar, 1995
71. Jiang et al., 1994
72. Knowles, 1977
73. Kong, 2018
74. Ladizinsky, 1995
75. Larsen, 2018
76. Leggett, 1985
77. Leggett, 1985
78. Majumdar et al., 1997
79. Mathre et al., 1985
80. Mercader, 2009
81. Montgomery, 2007
82. Nabukalu, 2018
83. Nadel, 2004

84. Nakasagga, 2018
85. Newell, 2018
86. Nurock 1968
87. Olsson, 2018
88. Oram, 1996
89. Paterson et al., 1995
90. Paterson, 2002
91. Paterson, 2018
92. Petrova, 1960
93. Picone, 2002
94. Piper, 1993
95. Pratt 1996
96. Purseglove, 1968
97. Reimann-Philipp, 1995
98. Roost, 1996
99. Rowlett, 1981
100. Sacks et al., 2000
101. Schaeffer, 1990
102. Schmit et al., 1996
103. Simmonds 1995
104. Smartt, 1995
105. Soost, 1996
106. Spiegel-Roy, 1975
107. Stutz, 1957
108. Turner, 2018
109. Van Tassel, 2002
110. Van Tassel, 2010
111. Wagoner, 1990
112. Wagoner, 1990a
113. Wagoner, 1995
114. Wang, 1985
115. Wang, 1990

116. Wang, 2018
117. Watkins 1995
118. Way et al., 1990
119. Weiss, 2004
120. Wetterstrom, 2004
121. Xiong et al., 1999
122. Zoeller, 2002
123. Zohary, 1975

7

Principles and Practice in Breeding Annual Horticulture Crops

CONTENTS

This chapter is about the annual crops. those which lives for the season and complete their life cycle and die. These crops are grown seasonally. With the increase in the population and their needs for food, breeders started to breed these annual verities to fulfill all the requirement of the increasing population.

This chapter explains all the annual crops varieties, their breeding methods. For example, roots and tuber crops like potatoes, sweet potatoes, carrot. These varieties further have several species that have been made for fulfilling the requirement of the consumer.

This not only benefit the consumers, but this also benefits the grower who is able to get enough amount of money for their crops. So, annual plants with their breeding varieties are explained in this chapter.

7.1. INTRODUCTION

The basic aim of an annual plant is to produce the seed in order to ensure the propagation of the future generations. It is believed to set the pretty flowers for attracting the insects, so that they can have pollination process.

This is the reason because of which deadheading or removing of spent flowers before the seed gets mature which induces the plant to set even when there are more buds and flowers, in the hope of generating a greater number of potential seed that can survive.

There are some hardy annuals, these annuals are believed to withstand with a little frost which is without being killed off and is considered to continue to bloom and can even set it for the next year. But it is believed that they do not carry indefinitely or usually die shortly after their second year begins. For example, Bachelor Buttons and Salvia Victoria.

Annual crops can be further divided into two more seasons, cool season annuals and warm season annuals. Even though it is true that they live for the entire growing season and many also not flower all the time. For example, pansies which will fade as the heat in summer increases, and Zinnias will not start flowering until the nights stays warm.

There are many vegetable and herbs those are annual season crops, like beans, cilantro, cucumber, and basils. Even there are most of the perennial vegetable those are gown as an annual, this is because they are only hardy in the warm climate.

Figure 7.1: Some of the varieties of annual crops.

Source: Image by Piqsels

These plants are required to produce the flowers and fruits continuously because of that they are harvested, not because they are allowed to go to seed. For example, tomatoes and eggplants. Annual flower is considered to bloom non-stop and especially when there is removal of the plant.

Growing of annual plants helps in keeping the garden in bloom in all the season. Annual flowers are also allowing to have garden every year.

7.2. COLE CROP

'Cole crops' are generally used to describe about several vegetables who belongs to mustard family. Cole crops basically includes vegetable like cabbage, broccoli, Brussels sprouts, cauliflower, kale, Chinese cabbage and knol-khol.

All the crops are considered as a cultivated varieties of the species *Brassica oleracea*. The Cole crops are believed to have one of the oldest histories of the European vegetables and are being widely grown throughout the world, these includes sub-tropical, tropical, and temperate regions.

Cauliflower and cabbage are considered as major vegetables in India and rest of them are considered as minor vegetable in India. Cole crops are basically rich source of vitamins A and C. these crops are considered to contain appreciable amount of minerals. For example, potash, sodium, calcium, phosphorus, and iron.

All these crops have been developed from *B. oleracea*, L. var. *sylvestris* , commonly known as wild cabbage, cliff cabbage or a Colewort's. These names are basically derived from 'Cole' crops. All these crops are having one common trait and that is there genetic potential to thicken various parts of the plants.

7.2.1. Cauliflower Breeding

Cauliflower botanically known as *Brassica oleracea* L. var. *botrytis L.* which is continually improved in order to increase the sustainability of the production, reduce the waste, and to enhance the nutritional quality.

For this purpose, it is necessary to develop the resistance against biotic stresses like insects and disease, and abiotic stresses like rainfall and high temperature. It is considered that breeding plays a vital role while addressing these issues which is through the development of a superior varieties or hybrids.

Figure 7.2: Varieties of cauliflower.

Source: Image by Pixabay

Cauliflower can be further divided into several subgroups, these subdivisions basically depends on the origin from Australia, Europe, India, and USA. Further, cauliflower is being divided into subgroups that are from the different region present within the European countries.

These countries are Netherland, Cornwall in the south west England, Germany, and Italy. Generally, Cauliflowers were earlier like the snowball type that is basically self-incompatible. It is believed that the Indian cauliflowers are the most early and are quite self-incompatible.

It is also observed that the Australian cauliflowers that are basically intermediate towards the late maturity are found to be quite self-compatible. The Australian cauliflowers are usually found to be very large. It is essential for carefully handling of the selected cauliflower plant when they are dug for the purpose of seed production.

Digging of a mature plant those with the root balls disturbed as little as possible, along with the removal of about half of the leaves, that is said to result in >95% survival rate in the Cornell University breeding programs.

It is believed that the curd will eventually start to elongate and even produce the flower stalks and plenty of flowers. If in any case, the seed crop produced is not found to be heavy, and with that the plant still remain healthy, then the plant will continue to develop the flowers from the undeveloped sections of the curd for several months.

There are some breeders who are found to be in trouble related to the saving field selections, and grafting of the curd portion which is on to the young stock plants that have been found to be successful and with that it does not delay the seed production.

Alternatively, the tissue cell culture has been found as a mean of a vegetative propagation. Small piece of the curd is being removed from the parent plant, surface is sterilized and small pieces of around 1 to 2 mm in the diameter are often placed on the paper wicks in the nutrient solution.

After this process, the roots and shoots get develop, and then they can be transferred into the pots and can be grown in order to produce the new plants (Crisp and Gray, 1975a, b, c).

7.2.2. Broccoli Breeding

Broccoli is basically heat tolerance, susceptible towards high temperature damage are considered as some of the major problems . This often makes the crop agronomically more suitable especially for the cultivation that is done in cool and moist climates.

Heat which is at the harvest time is considered not to be critical, except for the fact that high temperatures will reduce the period over which the crop is considered as marketable. It is considered that there is about 3 to 4 weeks prior to the head marketing which is basically a critical period when the growing point is slightly different in order to become a reproductive.

Bjorkman and Pearson in 1997, have studies about the development stage basically at which the temperature response is found to be most critical. It is considered to be most sensitive at the straightened and bowed stages, which is basically a short period of about 4 to 5 days. Leaf bracts can even enlarge and can also grow through the head.

Branching:

Branching also gets influenced by the temperature and as in the hot summers there is an increase in the degree of branching. It is found to be dominant on non-branching. In most of the cases there is extreme non-branching type of broccoli, the plant is found to be similar to the cauliflower in which there is no bud development in the leaf axils and if there is removal of head and then there is no lateral bud development.

Figure 7.3: Varieties of broccoli.

Source: Image by Pxfuel

Under these conditions, if the head is being removed then there will be no seed production on the saved plant. It is believed that the branching varies which depends on the cultivar and the temperature. It is found that the high temperature increases the degree of branching. The moderate branching is considered as desirable, but not the excessive branching.

7.2.3. Chinese Cabbage Breeding

Botanical name of Chinese cabbage is *B. rapa* which is basically a types of diverse *B. oleracea*. Chinese cabbage is found to be susceptible to almost all the pathogens, environmental stress, and pests those affecting the *B. oleracea*.

Figure 7.4: Varieties of Chinese cabbage.

Source: Image by Pxfuel

Telekar and Griggs in 1981 often discuss about the issues those are of specific importance in the breeding of the Chinese cabbage. Most types are found to be readily cross-compatible. In case *B. oleracea*, the hybrids have been made by using self-incompatibility, but recently there has been an increase in using the male sterility which are of both the genic and cytoplasmic types.

7.3. CUCURBITS

Cucurbits species includes muskmelons, cucumber, pumpkins, squash, and watermelons, who belongs to the family of Cucurbitaceae. Botanically it is considered to form a homogeneous group. While they are found to be extremely diverse in the vine and fruits characteristics. even the floral structure is found to be very similar in much of the cases.

Figure 7.5: Varieties of cucurbits.

Source: Image by Wikimedia Commons

Generally, the families are characterized by the flowering habits which is known as monecious. A monoecious species or variety is basically a one that bears pistillate or the female organ of reproduction. It is found to staminate or male organs in the separate flowers. Both the kinds of flowers are occurring on the same plant.

In the Cucurbitaceae, it is found that the male and female flowers are easily distinguishable, even before they open up. The showy corolla or the petals of the pistillate flowers are found to be attached to the end which often make it easy to be recognizable. For example, small undeveloped cucumber, melon, and squash.

Male flowers are found at the end of an ordinary flower stem, which is without any enlargement. The corolla is found to be surrounding the pistils or stamens which are respective sexes. Normal monoecious examples are squash, cucumber, and pumpkins, but there are certain varieties like watermelon and muskmelon those are believed to show certain modifications of this condition that are termed as and romonoecious.

Andromonoecious plants are basically the ones that bears a bisexual flower or a complete flower, instead of being pistillate; and in addition to purely staminate flowers. It is believed that at the cursory glance these are bisexual, hermaphrodite flowers, or complete flowers that appear like an ordinary pistillate.

The examination that takes place within the corolla will show that the presence is not only of the stigma but instead they are of functional stamens as well. Such flowers are found to be self-fertilized by their own pollens, or by the pollen those are from the purely staminate flower or from another flower present on the same plant.

7.3.1. Squash and Pumpkins – *Cucurbita ssp.*

Cucurbita or a yellow flowered cucurbit is considered as one of the most morphological variable genera which is present in the entire plant kingdom (Robinson et al., 1976). Around 22 wild and 5 are cultivated species those are found to be extremely diverse in the fruit color, shape, and size.

Then the cultivated species are reproductively isolate from each other by using the genetic barriers and even they can be identified by using the morphological characteristic (Whitaker & Bemis, 1964; Whitaker & Bemis, 1975; Nee, 1990).

The presence of constant and relatively high chromosome number i.e., 2n = 40, as well as the presence of complex isozyme pattern often suggest an allopolyploid origin which is for the genus (Singh, 1979; Kirkpatrick et al., 1985).

Archeological record of the New world often suggests that the Cucurbita was considered as one of the first plant that is domesticated (Nee, 1990). It is considered that Cucurbita- corn- bean is a complex that is forming the nutritional basis, for the pre-Columbian civilization in the Western Hemisphere (Whitaker & Bemis, 1975).

C. pepo. was considered as one of first species that is to be domesticated in the New World. Cultivation is done by the inhabitants of the Guila Naquitz that are caved and dated between 10,000 to 8,000 before present, corn, and beans by more than 4,000 years (Smith, 1997).

7.3.2. Cucumber - *Cucumis sativus*

In the recent comprehensive biosystematics monograph, which was by Kirkbride in1993, genus Cucumis is considered to include about 32 annual and perennial species that is divided into two distinct group which are defined by the geographic origin and chromosome number. In African species 2n = 24, and in Asiatic species 2n = 14 chromosomes.

African groups are considered to include melon (botanical name *C. melo*) and in the Asiatic group includes cucumber (*C. sativus*) and its probably an ancestor of *C. sativus* var. *hardwikii* (Royle) or simply *C. hardwickii* (Perltreves & Galun, 1985).

Studies those based on isozymes, restriction fragment length polymorphism, and chloroplast DNA are considered to support the distinction that is between the cucumber and melon (Perl-Treves & Galun, 1985; Perl-Treves et al., 1985).

There are basically two theories that are attempting to explain the existence of a haploid chromosome number n = 7 and closely related with the haploid chromosome number n = 12. The haploid chromosome number have a chance to increase to n = 12 by fragmentation or get even reduced n = 7 by fusion process.

Fusion is being supported by the increased amount of heterochromatin content in *C. sativus* (Ramachandran & Narayan, 1985) and is by the presence of chromosomal alterations in the *Cucumis* speciation (Sing & Roy, 1974).

Comparative genomics are between the *C. melo* and *C. sativus* that may also clarify the phylogeny of these species (Danin-Poleg et al., 2001).

7.3.3. Melon – *Cucumis melo*

African group or melon group is considered to have 30 species which is divided into six subgroups (Kirkbride, 1993). Melon and other 2n = 24 species that were originally said to be distributed across a large part of Africa and Middle East up to South Arabia and Pakistan.

However, there are some species that are also occurring in the Asiatic group range (Kroon et al., 1979; Ramachandran & Narayan, 1985). This is particularly in case of *C. hystrix* Chakr, which is only one who have number n = 24 native to Asia.

These species are of some particular interest, because of the morphological and biochemical characteristics that are similar to the *C. sativus* and chromosome number is equal to *C. melo*, which basically indicates a possible bridge that is between the two species (Chen & Adelberg, 2000).

7.3.4. Watermelon – *Citrullus lanatus*

The genus *Citrullus* is considered to consist of eight species and sub species. Watermelon is believed to be the only cultivated species of the genus that is diploid with 22 chromosomes where, 2n = 22 (Mallick & Masui, 1986).

Figure 7.6: Varieties of watermelons.

Source: Image by Flickr

Bitter fruit is considered as a watermelon ancestor that formed from the C. vulgaris Schrader (Mohr, 1986). Watermelon is basically said to originate in Africa and India (Mallick & Masui, 1986). Watermelon is an important crop present in the warmer part of Russia and other parts of Asia Minors, which is Near East, Japan, and China.

The cultivation of watermelon often began in the new world in the Massachusetts as early as 1629 (Mohr, 1986). Watermelons are considered to be bought by the Spanish to America and are considered to quickly became as a popular crop (Robinson & Decker-Walters, 1997).

7.3.5. Bottle gourd – *Lagenaria siceraria*

There are total six species those have been recognized as it belongs to the genus Lagenaria or to the white flowered gourds. One of them is the domesticated monoecious species *L. siceraria*, whereas other five are wild perennial, dioecious those are formed from Africa and Madagascar.

It is found that the basic haploid chromosome number in the genus is 11 where number 2n = 22 (Sing, 1990). Bottle gourd was being domesticated in Asia and at the same time it is found to be indigenous to Africa (Whitaker & Davis, 1962).

The tropical Africa often remains as a primary gene pool for these species (Sing, 1990). Bottle gourd was considered as the most widely distributed plant in the world (Heiser, 1979) which is with the long history of use in the old and new world (Robinson & Decker-Walters, 1997).

The bottle gourd cultivation is the old world was being traced back over 5,000 years BP (Robinson & Decker-Walters, 1997). The archaeological evidence has shown that the bottle gourd was being cultivated in the area of North America in 10,000 to 7,500 years BP and in the South America it is 6,000 to 5,000 years BP.

It is believed that there is no secure argument that can be even use for resolving the unusual bi-hemispheric distribution of the bottle gourd. The experimental evidences are suggesting that the reason behind the early spread from Africa to the New world can occur through the oceanic drift (Whitaker & Davis, 1962).

Presently, it is being cultivated throughout the subtropical and tropical regions of the world especially for the food and for useful gourds (Whitaker & Davis, 1962).

7.3.6. Loofah – Luffa ssp.

The genus luffa is being comprised of seven species. In which four of them are well differentiated species those are from the old world (*L. echinata Roxb.*, *L. acutangula*, *L. aegyptiaca*, and *L. graveolens Roxb.*) and the rest three species are from the new world (*L. quinquefida* (Hook. & Arn.) Seem., *L. operculata*

All the species are considered to have 26 chromosomes with 2n = 26 (Dutt & Roy, 1990; Heiser & Schilling, 1990). Early spread of the genus luffa was considered to in the new and old world, but it is observed that they both have cultivated species that first originated (L.) Cogn., and *L. astorii* Svens.) (Heiser & Schilling, 1990) in India (Heiser & Schilling, 1990).

7.3.7. Chayote – *Sechium edule*

Chayote is considered as the only cultivated Cucurbit in the sub family of Sicyoideae. Sechium was considered as a monotypic, but now is observed to contain eight species (Newstrom, 1990). The cultivated chayote is found as diploid with the chromosome number 2n = 24.

This species and all the other wild varieties are found to be native to the new world (Whitaker & Davis, 1962). Chayote was found to be domesticated in Mexico and Guatemala which is in the pre-Columbian times that is either from the wild forms of the species or from the closest relative *S. compositum* (Sm.) Jeffe. (Newstrom, 1990).

Figure 7.7: Image representing varieties of chayote.

Source: Image by Pxfuel

After the colonization by America, chayote spreading rapidly towards all the tropical areas of the new world and it also became as a popular staple item in the diet for the people of the old world (Whitaker & Davis, 1962). Chayote is being cultivated throughout the tropical and subtropical regions of the world (Newstrom, 1990).

7.4. ROOTS AND TUBER CROPS

Root and tuber (R&T) crops are considered to comprise a botanically diverse group which is of starchy staples and is found to be critically important for the food and agriculture. There are more than 30 species of R&T that are grown in the world today. There are many who are playing important role in increasing the income of the generation.

The foremost among these are in terms of aggregate outputs and is being considered to have estimated value of the production are potato (*Solanum tuberosum* L.), yam (*Dioscorea spp.*), cassava (*Manihot esculenta* Crantz) and sweet potato (*Ipomoea batatas* L.).

Figure 7.8: Varieties of root crops.

Source: Image by Pxfuel

Potatoes and cassava are generally from the Andean center which is the origin of the cultivated plants. Whereas, sweet potato is basically from the Central America (Vavilov, 1992). Yam also includes several species that are basically originated in the South east Asia, tropical America, and West Africa.

It is believed that there are some species those moved from Africa to North and South America, and there are other who travelled from Asia to Africa (Asiedu et al., 1997; Hahn et al., 1987). Whereas, other prominent R&T crops often include taro, yam bean, cocoyam, those are additional Andean species like arracacha, oca, mashua and ulluco.

It is believed that nearly all the complexes are polyploids and are vegetatively propagating. Basically, there are four major crops of R&T

namely – cassava, potato, yam, and sweet potato. These crops are believed to occupy almost 50 million hectare land worldwide.

Farmers are often producing about 640 million metric ton of these crops on the annual basis, and it about 70% which are being harvested in the developing counties. The production of these four major crops that takes place in the developing countries is considered to have an estimated annual value of more than US$41 billion which is between 1955 and 1977, which is near about one – fourth the value of the major cereals (Scott et al., 2000).

7.4.1. Potatoes

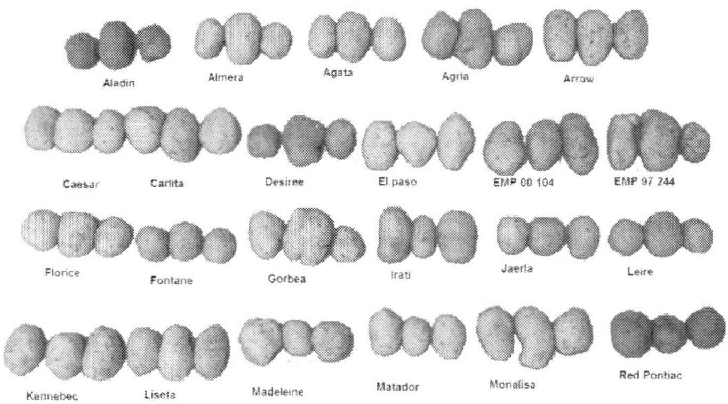

Figure 7.9: Varieties of potatoes.

Source: Image by Wikimedia Commons

Breeding Crop: It is considered that the somaclonal variation occurs when the plant regenerates and multiplication get involve in the tissue culture, which is particularly when there is a callus phase. After such variation it was first reported protoclones of the cultivars Russet Burbank (Shepard et al., 1980), it was further investigated as a breeding method which has its own rights.

There is desirable improvement that takes place on the parent's cultivars which was reported and assessed under the field conditions, but in most of the somaclonal variations it is comprised of undesirable change that basically did not merit further breeding efforts (Kumar, 1994; Veilleux, 2005).

Therefore, nowadays the somaclonal variation is viewed as the source of the undesirable variants that basically need to be screen out of the breeding programs which involves plant regeneration. In the same way, even though the natural mutation is considered as the ultimate source of all the genetic variations, the potatoes are not considered to be ideal for deliberating the mutation breeding because of its clonally propagated tetraploid crop.

Potato transformation was done by using the Agrobacterium mediated systems. This system was developed during the 1980s, first one was with *A. rhizogenes* (Ooms et al., 1986) and then the second one which was more successful and was with the *A. tumefaciens* (Stiekema et al., 1988; Dale and Hampson, 1995).

The gene of interest is being incorporated into the bacterial plasmid which is along with the selectable and promoter markers which are resistance toward antibiotic or herbicide, regeneration of the shoots with the selectable markers that can take place in the plant tissue culture which is basically in the presence of the selectable agents and the bacterium that can be co-cultured with the freshly cut tuber discs or leaf or internode explant of potato.

Recently, it is noticed that Chang et al. (2002) and Morris et al. (2006) are able to co-transform the potatoes simultaneously with the two gene by using the single selectable marker. The transgene constructs were further multiplied separately in the *A. tumefaciens* clones and are considered as a culture of Agrobacterium which was mixed and is being incubated with the potato internode explants.

7.4.2. Cassava

Cassava (*Manihot esculenta,* Crantz) is considered as the most important crop among the tropical root and the tuber crop (Pujol et al., 2002; Meireles da Silva et al., 2003). Along with the sugarcane (*Saccharum spp.*), maize (*Zea mays* L.), and rice (*Oryza sativa* L.), cassava is considered as among the most important sources of energy that is used in diet by most of the tropical countries of the world.

It is considered that the conventional plant breeding is one of the highest rates of return among the investments in the agriculture research. There has been remarkable increase in the productivity of many of the cassava crops which is during the twentieth century and was even reported due to genetic gains that are achieved through the process of crop breeding (Fehr, 1987).

Figure 7.10: Image of cassava crop.

Source: Image by Wikimedia Commons

Cassava is also considered as a benefit from the technology inputs especially in the area of breeding (Kawano, 2003). The new varieties of cassava in Asia, Africa, and LAC have satisfied all the needs of the farmers, consumers, and processors, by bringing millions of dollars which is in additional to the income to small farmers.

It is believed that the application of tissue culture technology made a positive contribution (De Vries and Toenniessen, 2001), and with that provides definition of adequate cultural practices that are particularly in relation with the fertilization protocols (Howeler, 2007).

Molecular biology and the genetic transformations (Blair et al., 2007; Calderón-Urrea, 1988; Fregene et al., 1997, 2000; Puonti-Kaerlas et al., 1997) had offered a great potential but they do not have any measurable commercial impact.

Cassava is considered as a major food crop and it is considered to have scientific breeding which began recently when compared with the other crops. There has been significant process that have been achieved particularly in Asia, where the use of cassava is mainly for the industrial process and with that it does not have major biotic constraints affecting its productivity.

Cassava breeding mainly faces several limitations that is important to be addressed. There is a use of heterozygous nature of the crop and the parental lines for generating the new segregating progeny that often makes it difficult to identify the parents with good breeding values.

Breeding is considered to be mainly based on the selection of the mass phenotypic recurrent. It is considered that there is very little knowledge on the inheritance of the traits which are of the agronomic relevance. There are several approaches that have been taken in order to overcome the constraints which is in the current methodologies for the genetic improvement of cassava.

Evaluation that takes place at the stages of the selection that allows for the estimates of the general combining ability that affects the breeding values of the parental lines. Inbreeding which is done by sequential self-pollination often facilitates the identification of the useful recessive traits those are either already found in the Manihot gene pool or is being induced by the mutagenesis.

7.4.3. Sweet Potato

Ipomoea batatas (L.) Lam. (Convolvulaceae, Dicotyledons) are being considered to produce the storage roots that are found to be rich in the carbohydrates and b - carotene, which is a precursor of vitamin A, and is believed to have leaves which are rich in proteins.

The roots are considered to contain vitamin C, B complex, and E as well as potassium, iron, and calcium. Purple fleshed sweet potatoes are containing the antioxidants like anthocyanins. The sweet potatoes are ranking seventh position in the world crop statics, which is just after the cassava with an annual production of about 9Mt and contains the cultivated area of about 110 Mha (FAO , 2009).

In most of the developing countries, it is considered as a smallholder crop tolerant which is of the wide range of edaphic and climate conditions and is basically grown with the limited inputs. There are basically two major difficulties that have been found to encounter in the development of the scientific breeding method.

Firstly, the sweet potato was considered as a difficult species for the Mendelian genetics because of the polyploidy. the segregation ratios are quite complex unless the single dominant allele is often involved. Secondly, the sweet potato is almost found to be self-incompatible.

Further, the sterility problem is common, which is a result of the high polyploidy. The practical consequences of the self-incompatibility and sterility are being recognized by the different researchers present in different countries (Martin, 1988; Wilson et al., 1989; Mihovilovich, et al., 2000).

It is found to be very difficult to produce the seeds by using self-pollination and by using the hand pollination which cannot produce more than four seeds and are often considered to have only one or two seeds per capsule.

Figure 7.11: Variety of sweet potato.

Source: Image by UIHere

Therefore, despite of having these practical constraints, which is one of the first modern breeding program that was implemented by the Louisiana State University in 1920s. from 1950s, the breeding programs have been considered to be implemented, which is mostly in case of temperate climates, in the United States, Japan, China, and areas of subtropical Taiwan (AVRDC).

7.5. TOMATO

Tomato botanically known as *Solanum lycopersicum* is considered as one of the important Solanaceous vegetable crops of the Peru-Ecuador origin

(Rick, 1969). It is being used as a fresh vegetable as well as a raw material for the products like ketchup, sauce, pure, juice, canned fruits, paste, etc.

Tomatoes are considered to have valuable source of the antioxidants or chemoprotective compound, apart from being having nutritive elements, flavour, and color to the diet. Tomato can be termed as "functional food" (Ranieri et al., 2004).

Figure 7.12: Varieties of tomato.

Source: Image by Pixabay

It is required to have desirable quality for tomato so that it can be used for processing which includes high total soluble solids of approximately 4 - 8° brix, pH should be less than 4.5, smooth surface, acidity should not be less than 0.4%, should be free from wrinkles, firm flesh, should have uniform red color, should have uniform ripening, and should have small core (Adsule et al., 1980).

The hybrid seed of tomato can be produced by using the following three methods:

- Using the male sterility and hand pollination,
- Using the hand emasculation and hand pollination methods, and
- Using the male sterility and natural pollination method.

7.5.1. Seed Production

(1) **Isolation:** Even though tomato is considered as a highly self-pollinating crop. It is necessary to provide an isolation distance of about 50 meters from other tomato variety or from the same variety that should not confer to the varietal purity requirements which is according to the seed certification laws. Isolation is being provided to avoid or eliminate any chance of outcrossing.

(2) **Land Requirement:** It is important that the tomato field should be free from the volunteer plants. Therefore, it is advisable that the plant seed crop should be sown in a field where previous crops was other than the tomato. However, there is no such specific condition specific for the seed certification standards regarding to the previous crop.

(3) **Field Inspection:** It is important to inspect the tomato seed field at least thrice during the crop season. This inspection is important to identify and remove the off-type plants. The first inspection was being made before the flowering and further the identification of the true-to-type plant which is ascertained on the basis of the foliage and characteristic of the plant type.

Then, second inspection is done at the time of flowering of fruiting stage. The flowers, fruits, and other plant characteristics form the basis of inspection. Then the final stage inspection is done on the color of the mature fruit, and its shoulders and the fruits shape which basically forms the basis for establishing the trueness of the type of the plants.

7.6. EGGPLANT

Brinjal eggplant botanically named as *Solanum melongena* L. which is a warm weather and is mostly found to be cultivated in the tropical and subtropical regions of the world. There are two other cultivated eggplant species namely, the gboma eggplants (*S. macrocarpon* L.) and the scarlet eggplant (*S. aethiopicum* L.).

These two varieties are less known, but it is considered that they have a local importance in the sub-Saharan Africa (Schippers, 2000; Daunay and Hazra, 2012). In case of the nutritional value of the eggplant, it is considered to have a very low caloric value and is also counted among the vegetable with the highest content of vitamins, bioactive compound, and minerals that are important for the human health (Raigón et al., 2008; Plazas et al., 2014b; Docimo et al., 2016).

It is believed that the eggplant ranks among the top 10 vegetables which is in terms of oxygen radical absorbance capacity (Cao et al., 1996). It is found that the bioactive properties of the eggplant are mostly associated with the high content present in the phenolic compounds (Plazas et al., 2013).

This means that mostly phenolic acids, especially chlorogenic acid which is in the fruit flesh (Stommel et al., 2015) and in the fruit skin its anthocyanins (Mennella et al., 2012). Here, both of them are considered to have multiple properties that are beneficial for the human health (Plazas et al., 2013; Braga et al., 2016).

7.6.1. Eggplant Germplasm in Breeding

The key issues of the breeding process are screening of the available accessions used for targeting the traits and morphological descriptions that are of the accessions. At the World Veg there is a large number of commercial cultivars, germplasm, and the landraces that have been examined in order to identify the desired genotype.

Figure 7.13: Varieties of eggplant.

Source: Image by Flickr

It has been especially used in the eggplant breeding program or for recommending towards the private sector seed companies and other partner institutions.

There is large variation in the yield parameters and in the fruit quality parameters that have been documented in the collection. These data have been combined over years and it can be further retrieved from the AVGRIS, which is the world vegetable center gene bank database system (2017).

One of the main objectives behind breeding *S. melongena* is that they are resistance towards the disease, insects, and nematodes. It is considered that the standardized descriptors often include the characters which are for both germination, inflorescence descriptors, fruits and seed traits, and vegetative phase, respectively. wilt (*Verticillium dahlia*), eggplant (fruit borer *Leucinodes orbonalis*), Bacterrial wilt (*Pseudomonas solanacearum*), Fusarium wilt (*Fusarium oxysporum*), leafhopper (*Amrasca biguttula*) , root-knot nematode (*Meloidogyne spp*.), melon aphid *Aphis gossypii*) etc.

It is found to be resistance towards the abiotic stresses like low or high temperatures, drought, and salinity reported to be parthenocarpy, resistance towards the herbicides. It is found that it has high nutritive value like sugar, total phenol contents, high dry matter, anthocyanin, low level of a polyphenol oxidase activity and ortho-dihydroxy phenolic compound in order to avoid the browning of the cut fruit.

The main S. melongena breeding objectives are:

- resistance to diseases, insects, and nematodes (Fusarium wilt Fusarium oxysporum,Verticilium wilt Verticillium dahliae,Bacterrial wilt Pseudomonas solanacearum,eggplant fruit borer Leucinodes orbonalis,melon aphid Aphis gossypii,leafhopper Amrasca biguttula, root-knot nematode Meloidogyne spp., etc.),
- resistance or tolerance to abiotic stress (drought, low or high temperatures, salinity),
- parthenocarpy,
- resistance to herbicides, resistance to diseases, insects, and nematodes (Fusarium wilt Fusarium oxysporum,Verticilium wilt Verticillium dahliae,Bacterrial wilt Pseudomonas solanacearum,eggplant fruit borer Leucinodes orbonalis,melon aphid Aphis gossypii,leafhopper Amrasca biguttula, root-knot nematode Meloidogyne spp., etc.)

7.7. CHILLI AND PEPPER

It is considered that there is a huge kind of pepper that are grown worldwide with different colors and shapes. The usage of pepper is considered as diverse as the types of capsicum fruits (Albrecht et al., 2012). Peppers are considered as an important source of the nutrients in the human diet (Shetty et al., 2013), and it is consumed dried or even fresh.

It is noticed that they promote the health benefits like diabetes and reducing the obesity (Vasconcelos, 2016). Pepper extracts are being used in the pharmaceuticals and cosmetics. Beside of using peppers as a spice, peppers have an ornamental potential when they are grown in the pot or gardens (Bosland and Votava, 2012).

The capsicum genus is said to contain about 30 wild species and also five domesticated species, namely *Capsicum annuum*, *C. chinense*, *C. frutescens*, *C. baccatum*, *C. pubescens* (Dewitt and Bosland, 1996; Arimboor et al., 2014; Sikora and Nowaczyk, 2014).

Botanically, it is considered that the chili peppers are being classified as a perennial especially when they are in the native habitats and the multiplication that takes place through seeds. Therefore, peppers are usually grown as an annual crop in the world. It is being noticed that frosting and low temperatures are hindering the field production (Votava et al., 2005; Dewitt and Bosland, 2009).

Flowers of the cultivated species are considered as hermaphrodites and even shows protogyny (Allard, 1960; Pickersgill, 1997). The fertilization system of the pepper is considered as a domesticated capsicum pepper that often performs the process of self-pollination.

But there is a possibility that due to environmental conditions it may also occur by cross-pollination (Villela et al., 2014). The frequency of the cross-pollination in the field is believed to range from just 2% to as high as 90% (Pickersgill, 1997).

Fruits that are formed can be hot or not, this particularly depend on each genotype. The pungency of the hot pepper is supposed to be because of the accumulation of the capsaicinoids, which is basically a group of alkaloids that are found to be unique to the genus of capsicum (Sun et al.,2015).

Capsaicinoids are being synthesized in the placental tissues which are localized inside of the fruits and in the same structure the seeds are adhered (Wahyuni et al., 2013).

7.7.1. Plant Breeding in Chili Peppers

One of the oldest domesticated crops in the Western Hemisphere would be the Capsicum pepper. Today, in various ethnic groups, it would be used as an essential spice across the world (Kim et al., 2014). Earlier to the arrival of Columbus in Americas (XVI century), chili peppers would be used by the natives, as food, as war artifacts and in religious rituals (Chiou et al., 2014).

Wild species of Capsicum pertaining to various regions of Mexico, Central and South America facilitated the occurrence of independent domestication process (Andrews, 1984). Assuming such wild species being a part of human diet since earlier to domestication (at least 7200 years earlier), not just in the form of a spice, but even as a source of food (Dewitt and Bosland, 2009).

As per the beliefs of ancient civilizations, peppers are considered to be laden with mystical and spiritual powers. In Aztec, Mayan, and Inca civilizations it was considered to have a lot of significance and during fasting fruits would be consumed for the purpose of flattering and pleasing the Gods (Bosland and Votava, 2012). The growing and selection of peppers are supposed to have been done for various reasons which definitely demonstrated the Native Americans as the world's first "pepper breeders".

Figure 7.14: Varieties of chili.

Source: Image by Wikimedia Commons

They are supposed to have been developing several varieties of peppers, cultivated presently. Certainly, the concept of plant breeding would be sluggish and casual majorly dependent on the skill of individual selection. Nonetheless, once considered as an art would have become "science" on the basis of principles proposed by Mendel related to genetics and heredity.

Presently, plant breeding (including chili peppers) would be involving several methods of selection which would be chiefly dependent on the planned goals and plants used as parents (Greenleaf, 1986; Singh et al., 2014).

Primarily, for gathering in a single cultivate, the strategy applied by plant breeders would be advanced genetic potential towards some characteristics, like productivity, resistance towards disease and bioactive compounds.

7.7.2. Plant Breeding Techniques in Capsicum:

Chili peppers would be considered to be diploid, largely performing self-pollination. Flowers would be perfect in capsicum genus with the male and female reproductive structures would be on the same flower (Allard, 1960). They would have close relation with potato, tomato, eggplant, tobacco, and petunia, also examples of Solanaceae.

The number of chromosomes (2n=2x=24) and domesticated species of peppers would be same for several members of the family, even though there would be a drastic difference in the size of genome pertinent to one genus to another. Few wild species would contain (2n=2x=26). Comparatively, the *C. annuum* genome (3.48GB) would be greater three to four time than the size of tomato (*Solanum lycopersicum*).

The length of average exon/intron would be 286.5 bp/541 bp; the gene would be almost 34.900; the length of transposable elements would be 2.34 Gb (76.4%) in total. Extremely conserved synthetic blocks would be shared by the hot peppers, along with the tomato genome, considered as its closest relative in the family of Solanaceae (Kim et al., 2014). There would be a record of familiar genes which could be valuable for pepper breeders.

The list initiated with 50 genes in 1965, these days it would be 292 various genes in total, like dw-1 (dwarf, plants with 15 to 20 cm in height), Ef (early flowering), me-2 (*Meloidogyne spp.* resistance) and others (Bosland and Votava, 2006).

The ones used more would be mass selection, pedigree method (or Genealogical method), single seed descent – SSD method, backcross, recurrent selection, and hybridization (Coon et al., 2008; Kulkarni and Phalke, 2009; Nsabiyera et al., 2013; Manzur et al., 2014). The best chosen method or of their combination would be dependent primely on the inheritance type (monogenic, oligogenic, or polygenic) from the traits that would require improvement (Lee et al., 2013).

Figure 7.15: Varieties of capsicum.

Source: Image by Pxfuel

7.8. OKRA

Okra would be a seed propagated, hot weather crop, sensitive towards frost, low temperature (below 15°C), swamping and drought conditions. It would also be called as lady's finger, bhindi, gumbo, etc. its growth would be limited to the tropical and subtropical regions as well as warmer parts of temperate regions.

Figure 7.16: Image representing okra crop.

Source: Image by Pxfuel

Hot humid climate would be ideal for the crop to thrive. An average temperature of around 25°C with high relative humidity (65-85%) would be required for the best growth of the plant along with fruiting.

At temperatures ranging from 30-35°C seed germination would be quick. Germination would be slowed down below 25°C temperatures, whereas,the plant and fruit growth would be sluggish with temperatures higher than 42°C, resulting in dryness/desiccation and flower buds dropping.

The tender fruits are cooked as vegetables. In nutritive value of the fruits would be relatively good with 100 g of its consumable portion containing dry matter of 10.4 g, energy of 3100 calorie, protein 1.8 g, calcium of 90 mg, iron of 1.0 mg, carotene of 0.1 mg thiamin of 0.07 mg, riboflavin and niacin of 0.08 mg and vitamin C of 18 mg with almost related constituents, excluding some, in leaves (Grubben, 1977).

A form of *Abelmoschus ssp.* would be used as leafy vegetable, in the countries of Far-East. 12% of edible oil and 20-24% of edible protein would be contained in the dry seed kernels of okra and might be used as an edible oil substitute. For off-season consumption, dehydrated and frozen fruit would be marketed.

7.8.1. Development of Hybrid

The developmental process of hybrids in okra would not be variant / different from other vegetables crops, viz., inbred development, combining ability testing and F_1 hybrid seeds production. In this case, breeding lines/cultigens/ varieties consistent for phenotypic characters could be regarded as parents for studying the combining ability and hybrid development.

It should be remembered that achieving 100% homozygosity in parents would be tough due to high number of chromosome and polygenic control of major economic characters. Using heterozygosity as such and favorable genes pooling in F_1 would be both crucial in okra.

Not just that the yield and its associated factors would require nurturing, but there might be several other goals not directly being related to the yield. For example, higher plant density and higher yield would provide upright unbranched plant.

Less pathogens and pests would be sheltered through deep cut, smooth leaves. Higher portion of plant dry matter would be provided by short internodes would be forming the economic part as fruit. Moreover, attractive color, shape and smoothness of the fruits might be favorable combination in the hybrids.

In few areas, several biotic and antibiotic stresses would limit production. It might be YVM virus, fruit and shoot borer, wilt, powdery mildew or conditions of soil salinity. F_1 hybrid incorporated with resistance or tolerance towards one or more such conditions would help in achieving continuous yield during challenging environment as well. Fetching the F_1 hybrid with thermo and photo insensitivity along with short day breeding types the crop yield would be for longer time.

Medium to late photosensitivity would be a characteristic for majority of the varieties (except Pusa Sawani). Significant reduction in yield is caused by moisture stress in most of the improved varieties. Tolerance towards salts, for germination and vegetative growth till 6 mmhos/cm NaC1 and 8 mmhos/cm Na2SO4 would be shown by the earlier Pusa Sawani variety.

7.8.2. Maintenance of Inbred/Pure Lines

Multiplication of pure lines/inbreds could be done from crossable forms in the isolated fields at 400 m. An area where okra would not have grown for three consecutive years in the past, maintenance is supposed to be done to prevent unwanted germinating plants contamination. Complete details pertinent to the specific characters of such line is supposed to be maintained.

Undesirable plant rogueing would be conducted at least at three stages – prior to flowering depending on the vegetative characters, during flowering on the basis of floral characters and at stages of fruit ripening. There would be many more suggestions for rogueing inspections. To keep all the genes together, there would be possibilities of allowing naturally inter-crossing or through mixed pollen within population.

Hosts of YVM virus should be kept away from plots and surroundings until 60% of fruiting is over. The YVM affected plants are required to be removed continuously. A season that would be most suitable for plant growth and fruiting, would require such multiplications, so that it would provide longevity and the greatest number of fruits on plants.

7.9. CONCLUSION

Annual crops are basically those crops that completes their lifecycle, i.e., from germination to production of seed which is within one growing season and then they die. Some examples of these types of crops are Cole crops, root and tuber crops, eggplants, and tomatoes. Likewise, there are many crops that are grown around the world seasonally.

There are many varieties of sub-species in these types of crops that are considered as an annual crop. Like, Cole crops include cauliflower, broccoli, Chinese cabbage, kale, and leafy vegetables, etc. Similarly, cucurbits also contain sub species like melons, watermelons, bottle guard, chayote etc.

Roots and tuber contain species like cassava, potato, carrot, sweet potatoes, etc. As the time passing, population of the world is increasing. With the increase in their population there is increase in their need for food.

If the traditional methods are being used for feeding this much of population, then there are chances that many of them will not be able to get sufficient amount of food. So, in order to provide them good quality of food and in a sufficient amount, it was considered to be necessary for crop breeding.

Crop breeding method basically helps in boosting the quality, quantity and help in fulfilling all the needs. So, there have been many changes in crop and these changes helped in feeding the food to the large population.

Like in the cucurbit family the crops are like pumpkins, melons, watermelons, bottle gourd, etc. these species have been changed according to the need and requirement and currently they are having several varieties of their own.

Sometimes, the crop is found as not suitable for all the types of environment, particular place but the people want to consume it. Then, this crop breeding methods help to bring changes in the crop and help to fulfill the requirement of the population.

This chapter is focusing on all such annual crops that are grown in the world and according to their requirement how the process of crop breeding done on those crops.

Okra, also known as lady's finger, bhindi, gumbo, etc., is a seed propagated hot weather crop sensitive to frost, low temperature (below 15°C), water-logging as well as drought conditions. It is grown in tropical and sub-tropical regions and also in warmer parts of temperate regions. The crop thrives well under hot humid climate. Best plant growth and fruiting is observed at around 25°C average temperatures with high relative humidity (65-85%). Seed germination is fast at 30-35°C temperatures. Temperatures lower than 25°C slow down germination while above 42°C slow down plant and fruit growth and cause drying/desiccation and dropping of flower buds. The immature fruits are cooked as vegetable. The fruits are fairly good in nutritive value and 100 g consumable portion contains 10.4 g dry matter, 3100 calorie energy, 1.8 g protein, 90 mg calcium, 1.0 mg iron, 0.1 mg carotene, 0.07 mg thiamin, 0.08 mg riboflavin and niacin, and 18 mg vitamin C, with almost comparable constituents, barring few, in leaves (Grubben, 1977). In Far-East countries, a form of Abelmoschus ssp. is used as leafy vegetable. Dry fruit shell and stem are used in the manufacture of paper and cardboard. Dry seed kernels of okra contain 13-22% edible oil and 20-24% edible protein and may be used as a substitute for edible oil. Dehydrated and frozen fruits are also marketed for off-season consumption. Brinjal eggplant (Solanum melongena L.) is an agronomically important non- tuberous crop of Solanaceae family, native to southern India, and widely grown in Americas, Europe, and Asia. Its close relatives: S. aethiopicum (scarlet eggplant) and S. macrocarpon (Gboma eggplant) are of African origin. Brinjal eggplant is a new vegetable in Poland. Until recently it was grown only in glasshouse conditions. Breeding of new eggplant cultivars and improvement of growing technology resulted in the increase of both cultivation acreage and consumption scale in Poland. For an efficient development of eggplant breeding programmes it is necessary to collect information concerning the characteristics and variety of eggplant genetic resources. The present publication includes detailed taxonomy of eggplant and its relatives, their origin and history of growing. The information connected with eggplant worldwide genetic resources and breeding objectives have been presented. Different propositions for cultivar grouping and the characteristics of selected cultivars have been taken into consideration

REFERENCES

1. Bisognin, D., 2002. Origin and evolution of cultivated cucurbits. *Ciência Rural*, [online] 32(4), pp.715-723. Available at: <https://www.researchgate.net/publication/26364541_Origin_and_evolution_of_cultivated_cucurbits> [Accessed 13 June 2020].
2. Ceballos, H., Iglesias, C., Pérez, J. and Dixon, A., 2004. Cassava breeding: opportunities and challenges. *Plant Molecular Biology*, [online] 56(4), pp.503-516. Available at: <https://pubmed.ncbi.nlm.nih.gov/15630615/> [Accessed 13 June 2020].
3. Cheema, D. and Dhaliwal, M., 2005. Hybrid Tomato Breeding. *Journal of New Seeds*, [online] 6(2-3), pp.1-14. Available at: <https://www.researchgate.net/publication/254370235_Hybrid_Tomato_Breeding> [Accessed 13 June 2020].
4. Dhaliwal, M., 2017. *Cole Crops*. 3rd ed. [ebook] Kalyani Publishers, pp.148-176. Available at: <https://www.researchgate.net/publication/313098865_Cole_crops> [Accessed 13 June 2020].
5. Dhankhar, B. and Mishra, J., 2005. Objectives of Okra Breeding. *Journal of New Seeds*, [online] 6(2-3), pp.195-209. Available at: <https://www.researchgate.net/publication/241746412_Objectives_of_Okra_Breeding> [Accessed 13 June 2020].
6. Dixon, G., 2007. *BREEDING, GENETICS AND MODELS*. [ebook] Available at: <https://www.cabi.org/ISC/FullTextPDF/2007/20073006211.pdf> [Accessed 13 June 2020].
7. Hundal, J. and Dhall, R., 2005. Breeding for Hybrid Hot Pepper. *Journal of New Seeds*, [online] 6(2-3), pp.31-50. Available at: <https://www.researchgate.net/publication/254370237_Breeding_for_Hybrid_Hot_Pepper> [Accessed 13 June 2020].
8. Iannotti, M., 2019. *What Is an Annual Plant?* [online] The Spruce. Available at: <https://www.thespruce.com/what-is-an-annual-plant-1401924> [Accessed 13 June 2020].
9. Mba, C., Li, X., Bonierbale, M. and Simon, R., 2003. *Genomics and Molecular Breeding for Root and Tuber Crop Improvement*. [ebook] Blackwell Publishing. Available at: <https://www.researchgate.net/publication/230719359_Genomics_and_Molecular_Breeding_for_Root_and_Tuber_Crop_Improvement> [Accessed 13 June 2020].
10. Padilha, H. and Barbieri, R., 2016. *Plant Breeding of Chili Peppers (Capsicum,Solanaceae) - A Review*. [ebook] Brazil: Australian Journal

of Basic and Applied Sciences, pp.148-154. Available at: <https://www.researchgate.net/publication/311639095_Plant_breeding_of_chili_peppers_Capsicum_Solanaceae_-_A_review> [Accessed 13 June 2020].

11. Shende, V., Seth, T., Mukherjee, S. and Chattopadhyay, A., 2012. *Breeding Tomato (Solanum Lycopersicum L.) For Higher Productivity and Better Processing Qualities*. [ebook] SABRAO journal of breeding and genetics. Available at: <https://www.researchgate.net/publication/281425285_Breeding_tomato_Solanum_Lycopersicum_L_for_higher_productivity_and_better_processing_qualities> [Accessed 13 June 2020].
12. Singh, B., Singh, B. and Singh, P., 2017. Breeding Cauliflower: A Review. *International Journal of Vegetable Science*, [online] 24(1), pp.58-84. Available at: <https://www.tandfonline.com/doi/abs/10.1080/19315260.2017.1354242?src=recsys&journalCode=wijv20> [Accessed 13 June 2020].
13. Taher, D., Solberg, S., Prohens, J., Chou, Y., Rakha, M. and Wu, T., 2017. World Vegetable Center Eggplant Collection: Origin, Composition, Seed Dissemination and Utilization in Breeding. *Frontiers in Plant Science*, [online] 8. Available at: <https://www.ncbi.nlm.nih.gov/pmc/articles/PMC5609569/> [Accessed 13 June 2020].
14. WHITAKER, T. and JAGGER, I., n.d. *BREEDING AND IMPROVEMENT OF CUCURBITS*. [ebook] Available at: <https://naldc.nal.usda.gov/download/IND43893559/PDF> [Accessed 13 June 2020].
15. Adsule et al., 1980
16. Albrecht et al., 2012
17. Allard, 1960
18. Andrews, 1984
19. Arimboor et al., 2014
20. Asiedu et al., 1997
21. Bjorkman and Pearson, 1997
22. Blair et al., 2007
23. Bosland and Votava, 2006
24. Braga et al., 2016
25. Calderón-Urrea, 1988

26. Cao et al., 1996
27. Chang et al., 2002
28. Chiou et al., 2014
29. Coon et al., 2008
30. Crisp and Gray, 1975a, b, c
31. Dale and Hampson, 1995
32. Danin-Poleg et al., 2001
33. Daunay and Hazra, 2012
34. DeVries and Toenniessen, 2001
35. Dewitt and Bosland, 2009
36. Docimo et al., 2016
37. Dutt & Roy, 1990
38. Fao, 2009
39. Fehr, 1987
40. Fregene et al., 1997, 2000
41. Greenleaf, 1986
42. Grubben, 1977
43. Hahn et al., 1987
44. Heiser & Schilling, 1990
45. Heiser, 1979
46. Howeler, 2007
47. Kawano, 2003
48. Kim et al., 2014
49. Kirkbride, 1993
50. Kirkpatrick et al., 1985
51. Kulkarni and Phalke, 2009
52. Kumar, 1994
53. Lee et al., 2013
54. Mallick & Masui, 1986
55. Manzur et al., 2014
56. Martin, 1988
57. Meireles da Silva et al., 2003
58. Mennella et al., 2012

59. Mihovilovich, et al., 2000
60. Mohr, 1986
61. Morris et al., 2006
62. Nee, 1990
63. Newstrom, 1990
64. Nsabiyera et al., 2013
65. Ooms et al., 1986
66. Perltreves & Galun, 1985
67. Perl-Treves & Galun, 1985
68. Perl-Treves et al., 1985
69. Pickersgill, 1997
70. Pickersgill, 1997
71. Plazas et al., 2014b
72. Pujol et al., 2002
73. Puonti-Kaerlas et al., 1997
74. Raigón et al., 2008
75. Ramachandran & Narayan, 1985
76. Ranieri et al., 2004
77. Rick, 1969
78. Robinson & Decker-Walters, 1997
79. Robinson et al., 1976
80. Schippers, 2000
81. Scott et al., 2000
82. Shepard et al., 1980
83. Shetty et al., 2013
84. Sikora and Nowaczyk, 2014
85. Sing & Roy, 1974
86. Sing, 1990
87. Singh et al., 2014
88. Singh, 1979
89. Smith, 1997
90. Stiekema et al., 1988
91. Stommel et al., 2015

92. Telekar and Griggs, 1981
93. Vasconcelos, 2016
94. Vavilov, 1992
95. Veilleux, 2005
96. Villela et al., 2014
97. Votava et al., 2005
98. Wahyuni et al., 2013
99. Whitaker & Bemis, 1975
100. Whitaker & Davis, 1962
101. Wilson et al., 1989

8

Transgenic Horticultural Crops

CONTENTS

Transgenic crops in horticulture brings the favorable genes in the plants, desired to be produced by the farmers. This is done by gene transfer from the same plant or from other organism to the plant. This chapter discusses in its first part, the meaning of transgenic crops, its uses, and the making of transgenic crops.

The meaning of transgenic plants along with its development and types are also explained in the first part. The middle of the chapter discusses about the advantages of transgenic crops. The disadvantages are also stated in this part.

The chapter also discusses about the transgenic in ornamental plants, the use of biotechnology in ornamental crops. Genetic modification of the crops and major genetically modified ornamental crops are discussed. The chapter ends with the topics such as genetic transformation in fruits and nuts and the assessment of risk of the transgenic crops.

8.1. INTRODUCTION

Earlier, before the introduction of transgenic in plants and crops, the crops were produced and altered through the method of selection. The method of selection has become a widely used method in the production of the crops. It has become a dedicated science in itself.

The introduction of the transgenic method has raised a question that 'why is there need for the use of genetic techniques (transgenics) for the improvement in crops over the recent decades?' In simple words, the selection method of producing the crops takes many generations (i.e. large investments of time) and also it does not always work.

With the use of transgenics, the crop production can be customized as having the desired traits. It also helps in increasing the yield that makes it easy for the crops to last for long time and grow without having affected from pest and other diseases.

The genetic technique, transgenic makes it easy to feed the growing needs of the population by producing the crops with desired traits.

8.1.1. What is a Transgenic Crop?

A transgenic crop is basically a genetically modified organism (GMO). By the word 'transgenic', it refers to the use of recombinant DNA technology by transferring the genes. Additionally, a crop which is grown with the help of transgenic technique, contains one or more genes. Those genes could be inserted artificially either from an unrelated plant or from different species altogether.

Figure 8.1: Genetically Modified Organism.

Source: Image by Wallpaper Flare

8.1.2. Why do we use Transgenic Crops?

One of the common purposes to produce the genetically modified plants is to deliver the best possible products. Over the years, these modifications are for the production of the best produce. Before the introduction of recombinant DNA technology, the improvements in the products were possible.

But, those improvements or changes in the crop were difficult. The improved product has to be sacrificed with the change in the color or the size of the plant to make it more appealing to the eyes of the buyer.

Practically, the crop can be made tolerant to the changing environment as to cold, frost, or drought that leads to the easy growing of the crop. These goals would not be achieved with the conventional method of selection programs. Therefore, the new technique preferred over the old method of selection is the transgenics, for the production of new products in order to improve sales.

The biggest challenge nowadays for the crop production is the modification of crops for increasing the resistance of plants to insects and diseases. The improvement in the resistance to these disease and insects reduces the need for the use of herbicides and pesticides.

By the use of this technology, the farmers and the consumers are benefitted and it helps them save their money on chemicals. Conclusively, the economically safer method to produce the crops is the method of transgenic production of crops that makes the crop appealing and potentially profitable.

8.1.3. How are Transgenic Crops Made?

For many years, it has been seen that the plant breeding is the selection of finest plants for getting the best crop. The variation in the plants at that time was done through induced mutation or hybridization in which two or more plants were crossed to get the desired result.

The method of selection was done through the nature which uses a "selection of the fittest" concept. According to the concept, only the seeds best adapted to that environment succeeded.

Such as, the farmers selected only those seeds whose size was the biggest with non-shattering seed heads. They assumed those seeds to be a best fit. In today's era, the scientists can select and also create crops by inserting genes to make the desired seed having any trait desired.

For the making of the transgenic crops, there are broadly five stages for the crop production, they are: extracting DNA, cloning a gene of interest, designing the gene for plant infiltration, transformation, and finally plant breeding.

In order to know about these stages, the knowledge about DNA (deoxyribonucleic acids) is necessary. Basically, DNA is the universal programming language of all cells and the genetic information is stored in this. There are thousands of genes that are there in a DNA. The genes are the discrete segments of DNA which encodes the information that are necessary for the production and assemble of specific proteins.

8.2. TRANSGENIC PLANTS

The plants whose DNA is modified using the genetic engineering techniques are the transgenic plants. The main objective of using the transgenic technique is to introduce a new trait to the plant which does not occur naturally in the species. A gene that has been inserted artificially is the transgenic plant.

The gene sequence in the plants is called as the transgene. This gene could be from an unrelated species or from a totally different plant. The objective being the production of best suited plants as soon as possible. The process of this gene transfer can be very useful such as improved quality, tolerant to heat, drought resistant, higher yield and improving the shelf life. It also saves the plants from a variety of biotic and abiotic stresses.

The production of transgenic plants can be done in such a way that it gives industrial and pharmaceutical value with that of foreign proteins.

Plantibodies are the plants that are made up of vaccines or antibodies. These plants are especially striking as these cannot get affected from any human disease. Thus, it helps in reducing the screening costs for viruses and bacterial toxins.

The year 1983, reported the first transgenic plant. It is since that time, many other recombinant proteins have been expressed in several important agronomic species of plants. The species includes tobacco, banana, tomato, potato, canola, alfalfa, and corn. Generally, tobacco plants were used and the potatoes and bananas were also considered for the purpose of vaccines for human beings.

8.2.1. Development of Transgenic Crops

Generally, the making of the genetically engineered plants is done in the laboratory. The plants are altered by giving the genetic touch-up and adding one or more genes of a plant's genome. The nucleus of the plant-cell is considered as the target for the new transgenic DNA.

Most of the plants are modified genetically using the method of biolistic or Particle gun method or by *Agrobacterium tumefaciens* mediated transformation method. The Biolistic method also known as "Gene Gun" method or "Micro-Projectile Bombardment method" is the most common method used in species such as corn and rice.

Under this method, the DNA is bound in the form of tiny particles of gold or tungsten which is shot y gun in high pressure into the plant tissue or single plant cells. The particles gunshot is penetrated both in the cell wall and the membranes.

The DNA separates from the coated metal, thereby integrating inside the nucleus into the plant genome. This method has worked best for the monocots and many other crops such as wheat or maize. However, the transformation by using the *Agrobacterium tumefaciens* is not much successful.

This method is having a disadvantage that it could damage the cellular tissue. However, the technique is considered as safe and clean. Another method which is used for developing the genetically engineered plants is the "Agrobacterium" method.

The method basically uses the soil-dwelling bacteria which is known as *Agrobacterium tumefaciens*. The *Agrobacterium tumefaciens* carries the ability to infect plant cells with its DNA.

Figure 8.2: Agrobacterium tumefaciens.

Source: Image by Flickr

A tumor (Ti plasmid) is used to induce a plasmid with the piece of DNA (that infects the plant) being integrated into the plant chromosome. The Ti plasmid can control the plant's cellular machinery and use it to make many copies of its own bacterial DNA.

A large circular DNA particle has the capability of replicating independently of the bacterial chromosome. This plasmid has an advantage that it has the regions of transfer DNA (tDNA). There can be insertion of the gene that can be later transferred to a plant cell through a process known as "floral dip".

The dipping of flowering plants into a solution of Agrobacterium that carries the gene of interest. This process being followed by the collection of the transgenic seeds directly from the plant. This is the process of floral dip. The process is useful and accepted worldwide as a natural process of transfer and hence taken as a more acceptable technique than any other.

Additionally, "Agrobacterium" has the capability to efficiently transfer large fragments of DNA. However, it has one big limitation that in Agrobacterium, all the important food crops cannot be infected by these bacteria. This method is well suitable for the dicotyledonous plants like tomatoes, potatoes, and tobacco plants.

8.2.2. Types

The genes are inserted in the transgenic plants. Those genes are inserted from another plant. The genes can be taken from the same plant i.e., from plant to plant or between the kingdom i.e., bacteria to plant. In some cases, there is a need to modify the DNA correctly and efficiently in order to express it correctly in the host organism.

Figure 8.3: Cisgenic plants after being genetically modified.

Source: Image by Flickr

Under the Cisgenic plants, the conventional plant breeding can happen. Cisgenic plants are made up of genes which are found within the same species or a closely related one. One of the advantages of using the modification through cisgenic is that it could be useful for plants that cannot be crossbreed by conventional means. For example, potatoes.

The plants that are modified using the cisgenic technique are not necessarily need the same level of legal regulation as any other genetically modified organisms would need.

8.3. ADVANTAGES OF TRANSGENIC PLANTS

The genetic modification technology is used extensively for producing a variety of crop plants till date. The population is growing day by day and hence food is the basic requirement for life that still remains a scare resource.

The food that is genetically engineered have much benefit to offer. It can improve the production yield, lowers the transportation cost, and enhance the nutritional content. In countries like USA and Canada, the development with regards to the quality of food they provide has set a benchmark for others as well as on conferring resistance to insect, pests or viruses and tolerance to specific herbicides.

The farmers are the ones that have been benefitted by these traits. However, other than these, there are no other benefit to the consumers. There are limited cases reported that a decrease in the prices of the products are due to reduced cost and increased ease of production. In the coming years, there are several genetically modified crops that are to be grown for malnutrition purpose.

8.3.1. Herbicide Resistant Plants

The herbicides resistant plants are those plants which can tolerate the herbicides. Many of the broad-spectrum herbicides has a special ingredient of glyphosate, which is the most active ingredient in this. By the transfer of are, the crops such as glyphosate resistant transgenic potato, tomato, tobacco, cotton etc. are developed.

The gene is transferred to the glyphosate EPSP synthetase from *Salmonella typhimurium*. The production of *E.coli* ulphonylurea resistant tobacco plants is done by transforming the mutant ALS (Amyotrophic lateral sclerosis) gene from Arabidopsis.

8.3.2. Insect Resistant Plants

There are numerous insect pests which are pathogenic for the bacterium known as, Bacillus thuringiensis. It eventually produces a toxin protein that mediates its lethal effect. With the help of recombinant DNA method, there can be direct introduction of the toxin gene into the genome of the plant. There is the expression that saves the plants from the insect pests.

8.3.3. Virus Resistant Plants

The production of Tobacco Mosaic Virus (TMV) resistant tobacco and tomato plants are done by the introduction of viral coat proteins. There are some other viral resistant transgenic plants such as Potato virus resistant potato plants, YMV resistant green gram, RSV resistant rice and YMV resistant black gram etc.

8.3.4. Pest Resistant Plants

If there is a resistance to a specific pest by a transgenic plant, it is a great benefit to the farmers. For example, in Hawaii, since 1996, there has been production and commercialization of the Papaya-ring-spot-virus resistant papaya. If the use of the pesticides is reduced, it will be a benefit to the environment.

The amount of insecticide has been reduced by the transgenic plants that contains insect resistance genes from *Bacillus thuringiensis*. However, there is adaptation of populations against pests and disease that causes organisms.

8.3.5. Nutritional Benefits

Many children in this world have become blind either partially or fully each year which is due to the lack of vitamin A deficiency. There is large human population in this world whose staple food is the millet rice.

There are three types of genes introduced by the researchers into rice; one from a microorganism and the other two from Daffodils.

As a precursor to vitamin A there is increase in the production of beta carotene in the transgenic rice. Bus seat is yellow in color. Therefore, the problem of vitamin A deficiency mostly in the young children can be treated by such useful tool of yellow and golden rice.

8.3.6. Use of Marginalized Land

There is marginalization of both coastal and terrestrial land in the vast land mass across the globe, due to excessive alkalinity and salinity. The other plants are given a salt tolerance gene from mangroves. It was identified cloned and transferred in other plants. In transgenic plants it was found that it is tolerant to higher concentrations of salt.

There is generation of salt tolerant transgenic maize plants for which gut D gene from *E. coli* was used. In order to set a good cropping system, such genes are helpful (MS Swaminathan, Personal Communication, 2000). At the University of California, Davis campus, the researchers have created transgenic tomatoes that grew in saline soils.

8.3.7. Reduced Environmental Impact

In this global world, one of the big global issues are the availability of water and its efficient usage. Also, the soil is subjected to extensive tillage known as Plowing. Because the soil gets eroded easily and it is difficult for controlling weeds and preparing seed beds, followed by serious loss of water content.

In the traditional communities, for many years the low tillage systems were used. Such crops are needed to be grown that can survive under such conditions. Also, there is need for the introduction of resistance to root diseases currently controlled by tillage and to herbicides can be used as a substitute for tillage.

8.3.8. Therapeutic Proteins from Transgenic Plants

With the use of recombinant DNA technology, the proteins of therapeutic importance can be produced in plants, like those used in the treatment and diagnosis of human diseases. There is industrial production of proteins though the scaling-up of these transgenic plants to fields.

> There is another field of research combining molecular Biotechnology and Agriculture called as Molecular farming or Bio pharming.

For the production of proteins in the transgenic plants for therapeutic use, these are of three types:

- Vaccines
- Proteins
- Antibodies

Antibodies fights against some diseases in the transgenic plants and these are directed against dental caries, rheumatoid arthritis, rhinovirus, malaria, cholera, *E. coli* , diarrhea, certain cancers, HIV, rhinovirus, influenza, hepatitis B virus and herpes simplex virus.

The plants like bananas and potato are produced with the vaccines which fight against infectious diseases of the gastro intestinal tract. Another potential target is cereal grain. In rice and wheat seed, there is an anti-cancer antibody that has recently been expressed.

It basically helps in recognition of cells of lung, colon, and breast cancer. Therefore, it is considered as useful in both diagnosis and therapy.

8.4. OTHER BENEFITS OF TRANSGENIC PLANTS

The genetically modified (GM) food has an enhanced ability that can withstand long-distance transportation. In order to give the transgenic plants a longer shelf life, these crops are picked and the green plants are allowed to ripen during transportation. The product reaches its destination without spoiling even after the prolonged shipping and storing periods.

As per the manufacturers of these genetically modified crops, the GM seeds usage provides with the number of benefits, having increased yields and decreased costs as the basic benefits of these crops. It has been said that the GM crops acts as a second "Green Revolution" in a world that fulfills the needs of billions of hungry people.

8.4.1. Improvement in Yield

The technology of genes plays an important role in increasing the productivity of food, fiber, and vegetable crops. The production of these crops ensures food security which is the basic thing for international peace and stability. There it is considered as an important means to fight for hunger.

Generally, the transgenes do not yield enhancing genes. By protecting the crops by insects, abiotic factors and other diseases, the yield of the crops can be increased. With the increasing world population, the gene technology is a potential factor to keep pace with the food production.

8.4.2. Improvement in Quality

The quality of any food is judged in three ways, such as market (keeping) quality, nutritional quality, and industrial quality. With the production of genetically modified crops, there is improvement in all these three types in different crops.

For example, with the use of gene technology on tomatoes, the transportation to distant places has been made easy and it can travel longer distance without ripening. This is done through the manipulation in the gene that helps in encoding the enzymes responsible for ripening (Ethylene Forming Enzymes, EFE) and softening (polygalactonase). Some of the other improvements in terms of the quality include non-browning potato, improved oil content in oilseed crops and starch composition of wheat flour.

8.4.3. Rapid and Accurate Technique

This method is considered as the rapid and highly accurate method of crop improvement. The cultivars can be developed by this technique which takes approx. 4-5 years. While, if it is produced through conventional (hybridization) method, it takes around 10-20 years. This is the reason that, this is a highly reliable technology.

8.4.4. No Barrier for Gene Transfer

There is no restriction on the transfer of gene between any two individuals whether related or unrelated. There can be transfer of gene from the same plant or from the micro-organisms to higher plants or even from animals to plants. Therefore, it overcomes all the barriers of gene transfer.

8.5. DISADVANTAGES OF TRANSGENIC CROPS

The production of transgenic crops was an issue for many years. There are many concerns regarding the GM crops and has been raised which are falling into the following two categories:

8.5.1. Effect on Human Health

The serious concern is regarding the adverse effect of the transgenic material on human health. For example, many people are allergic to transgenic crops, however, it is not certain that only the transgenic crops are the source of this allergic reaction.

Moreover, the antibiotic resistance genes are placed in these crops. These are suggested to be resistance to antibiotics that ultimately leads to super bugs. The bugs are not killed with antibiotic treatments.

8.5.2. Damage to the Natural Environment

Another concern is regarding the transgenic crops that may cause damage to the natural environment. For example, pollen from transgenic corn carries the capability to kill the monarch butterfly larvae. In the hybrid corn, there is a bacterial toxin in its pollen, which is afterwards dispersed over 60 m through wind.

The range where the pollen has been spread gets deposited on all the plants near cornfields. There it can be ingested by non-target organisms which includes the monarch butterfly ultimately leading to their death.

8.6. TRANSGENICS IN ORNAMENTAL CROPS

Horticulture is one of the growing industries and cultivation of flower crops is considered to be one of the lucrative and income generating sector. This sector involves ornamental plants that are very important part not only for this industry but also for human life because of its aesthetic and economic importance.

It is an important sector of horticulture industry and plays a major role in strengthening the economy of several African countries as flower crops are emerging as a major income generating commodities in Costa Rica, Colombia, Kenya, Ethiopia, and Ecuador.

The industry comprises of different varieties of ornamental plants and flowers which not only fulfill human needs but also aims at achieving sustainable development of the eco system. It comprises of cut flowers, loose flowers, pot plants, flowering and foliage, ornamental grasses, trees, shrubs, annuals, and other plants of ornamental value.

This industry has seen a significant growth in recent years and bringing new ideas and products as per the customers' requirement would ensure its continuous development. Innovation could include a new flower color, formation, fragrance, or a new hybrid flower crop.

This has not only been a factor for this industry to grow but also made it more competitive which consequently requires improvements in the existing products as well. The research work has doubled up to bring in innovation, F_1 hybrid breeding has been one of those popular method in flowers crops and breeders that have developed promising hybrids in ornamental crops that includes lawn grasses and cut foliage plants.

Despite of using various methods to bring novelty in the industry, there has been a need for alternative research programs as the continuous adoption of conventional breeding strategies in improvement programs, the available gene pool for developing new traits is becoming narrower, which limits their use in further improvement programmers.

For Example, in some sterile flower like orchids, hybridization or other conventional methods are not very useful and hybrid breeding and mutations are quite difficult to achieve, hence, the need for more methods for bringing novelty in ornamental plants was felt.

Floricultural crops are also developed asexually by vegetative reproduction, which is commercially important for ornamental species, but these crops are highly heterozygous which causes inheritance of several genetic factors.

This inheritance makes the improvement process for such crops more difficult during conventional breeding hence acting as a barrier in bringing innovation to the industry. One side it is difficult to find the gene of interest in natural gene pool and on the other side, the mix of favorable and unfavorable genes makes it more difficult to rely upon conventional breeding methods or hybridization. Also, it makes it harder to select genes for improvement of ornamental crops where a specific change or gene for a specific trait.

Another drawback of conventional breeding which arises the need for more developed and new research plans is that in such crops the sources of resistance for pests and diseases are limited, for which the need occurs to transfer the resistant genes from external sources like other species or unrelated organisms.

All these challenges and restrictions of conventional methods or it can be said that classical breeding specifically pertaining to ornamental crops necessitate the development of new methods for alternative breeding strategies that can be used to bring Novelty in the industry on continuous basis.

One of the recent developed and used method is transgenic breeding, the transgenic plants are those whose DNA is modifies using genetic engineering which introduces a new trait to the plant artificially.

It has worked very well on several crops like maize, soybean, canola, and cotton. It involves incorporation of herbicides and insect resistance which has definitely helped in making crops more resistive to pests and helped in overcoming the demerits of classical breeding by developing new traits in plants.

The new traits might be a unique color, resistance against harmful diseases and pests, which might be developed without altering the genetic structure of the plants and its usage. The precision and improvement of a trait without altering the genetic constitution of an elite genotype is the main advantage of transgenic breeding.

Another reason why the genetic engineering process is recommended for ornamental crops is that they are ideal product for it as the change in DNA / genes doesn't harm the population using it as it is not widely used for consumption and hence the chances of having any health hazards are very less unlike the other consumable products developed through transgenic process as they are for human consumption and people are worried about their safety standards.

However, due to varieties, increased profitability and ecological stability, many researchers are in favor of developing transgenic crops that are consumable directly by human population, which further provides scope for improvements in ornamental crops where researches are still underway.

8.6.1.Cut Flowers

The uses of cut flowers are for the fresh flower arrangement. These are cut from the plant with trimmed thorns, and it becomes ready to be used in the flower arrangements. Some of the common cut flowers are roses, lily, carnations, tulips, chrysanthemums, and gerberas.

Ornamental plants and grasses are allowed to grow to its fullest. Then, these plants are used in the landscape similarly as the perennials or other ornamental plants. The restios, sedges, and cat tails are some examples.

Lawn or turf grasses are the grasses which ensure that the ground is covered. The ground cover is essential in some places such as private lawns, sporting field and golf courses.

Figure 8.4: Lawn or turf grasses, that covers the surface.

Source: Image by Pexels

8.6.2. Potted and Indoor Plants

Potted plants are generally grown and decorated inside the residences. Generally, they are used for decoration purposes, health reasons and as an indoor air purifier. Some examples of such plants are Cacti, Ficus, Bonsais, Dracaena, Bromeliads, and Poinsettia.

Figure 8.5: Potted indoor plants.

Source: Image by Pexels

Bedding plants are those that are grown in quantity. These are generally grown in pots or flats in greenhouses. The plants are grown with the intent of transplanting it to a flower garden, window box, hanging basket, and other outdoor planters. Major names of the bedding plants are petunias, impatiens, and marigolds.

8.6.3. Trees and Shrubs

Trees and shrubs are generally used for gardens and landscaping. Some of the examples of ornamental trees are cedar, cherry blossoms, palm trees, and mulberry. The ornamental shrubs include Lavender, Ivy, Hibiscus, Magnolia, and Ficus species.

Figure 8.6: Cherry blossom tree used as an ornamental shrub.

Source: Image by Wikimedia Commons

8.7. USES OF BIOTECHNOLOGY IN ORNAMENTAL HORTICULTURE: CELLULAR ENGINEERING

The main focus of the Cellular engineering in ornamentals are on the micropropagation, haploid and polyploidy breeding.

8.7.1. Micropropagation

Micropropagation is used for multiplying the crops just like those which are genetically modified with the help of conventional plant breeding methods. This is one of the widely used methods by the Ornamental industry. They have used it for multiplying the plants on a large-scale with elite superior varieties. Micropropagation is considered as the most used approach for ornamental crops as the propagation by cloning is a fast process and it creates the exact plant as the mother plant.

8.7.2. Haploid Breeding

This method is helpful in the production of disease-free, haploid individuals (the last that have only half the number of chromosomes) can result in dwarfed ornamentals, for example, the Pelargonium variety "Kleine-Liebling" and the rose hybrid "Sonia".

The homozygous individuals upon chromosome doubling can easily be produced by haploid breeding, that accelerates the breeding process.

8.7.3. Polyploid Breeding

The use of polyploid individuals in ornamental horticulture is wide. They are having multiple sets of chromosomes. Basically, they are used for their improved characters such as larger flowers and thicker petals. One of the features of the polyploid individuals is that it can sometimes appear in nature spontaneously or can be induced by *in vitro* chromosome doubling.

8.8. GENETIC MODIFICATION

There are varieties of important food species that are being genetically modified for the development of its varieties in food species. Also, various other measures and efforts are being taken for the development of varieties in ornamental plants, though not at the same scale. It is especially done for flowering ornamentals.

However, there is difficulty in the development of these new varieties using the hybridization or mutagenesis methods as it is considered a lengthy and improbable process and especially if the varieties are completely sterile, for example as orchids.

Figure 8.7: Orchids having a sterile variety and being genetically modified.

Source: Image by Pixabay

Therefore, genetic modification has been evolved that gives answers to all the questions with regards to its variety improvement. The developmental time for the variety development in this industry is eased by biotechnology that shortens the duration of variety development. Here, the phenotypic novelty, such as flower color is considered as an attractive marketing factor.

There are several traits in ornamental plants that have already been modified. Such as fragrance, flower color, the shape of the flower, its architecture, postharvest life, flowering time, and resistance including both biotic and abiotic stresses.

Currently, there are 50 ornamental plants that exists and can be transformed. Numerous techniques are being used to produce these transgenic ornamental plants. The most common techniques used are Agrobacterium-mediated transformation and particle bombardment method.

The traits of the ornamental plant are classified on the basis of their value in the market chain. The person who grows these plants and its traits carries more value than to the consumer. The traits are based on how the production is and the shipping like, disease resistance and shelf life.

Moreover, the cost of these plants to consumer is high with respect to the other traits such as novel colors, modified growth, dwarfed plants, flower shapes, improved fragrance, and flower size.

8.9. MAJOR GENETICALLY MODIFIED ORNAMENTALS

8.9.1. Carnation

The genetically modified ornamental product that is on a commercial scale is the moon series from Suntory limited and Florigene private limited. There are release color of carnations is found in moon series that are commercially available in various parts of Australia Japan USA and European Union.

Figure 8.8: Varieties of carnations.

Source: Image by Wikimedia Commons

It was approved by Colombia in the early 2000s. Recently there is addition of four new carnations in the moon series. They are named as moonique (purple in color), moon berry light purple, moonpearl lavender, moon velvet dark purple.

They are added and approved for commercial use in 2012 in Malaysia. Carnations having longest vase life was developed, however it was not commercialized.

8.9.2. Chrysanthemum

Chrysanthemum has been recognized as one of the most important ornamental plants in the world. Currently molecular breeding for the blue Chrysanthemum is ongoing. In order to improve several other aspects of the flower the molecular technology is widely being used.

Figure 8.9: White Chrysanthemum.

Source: Image by Wikimedia Commons

The flower can be protected from insects, there can be modification in the color, abiotic stress resistance, altered plant architectures and pollination control can be achieved by using the transformation techniques. Currently there is no commercial availability of chrysanthemums.

8.9.3. Roses

One of the most popular transgenic flowers is the blue rose. There are three genes included in the transgenic rose, a delphinidin gene from blue pansy, a synthetic RNA interference gene to shut off the DFR gene, and a DFR gene from iris having an affinity for producing delphinidin.

Figure 8.10: One of the transgenic flowers is blue rose.

Source: Image by Pixabay

8.9.4. Petunia

Petunia is a flower which is commercially available. This flower contains an altered flower color which was developed by Beijing University. In Germany, an ornamental bioscience is mainly giving attention on specifically frost tolerance and an improved abiotic stress resistance flower. By this, it would help in increasing the range of environments in which petunia, a bedding plant could be grown.

Figure 8.11: Petunia flower.

Source: Image by Florist Durhum

8.10. GENETIC TRANSFORMATION OF FRUIT AND NUT SPECIES

Genetic transformation is considered as a power tool in plant breeding programs. It involves injection of a foreign gene into a plant cell that genetically transforms the plant and brings a new trait into it. Specifically, for heterozygous plants like tree fruit species which needs many years to mature and occupy a large place in growing hybrids.

The new traits that are developed among plants using genetic transformation could either be a code for diseases or pest resistance, herbicide resistance or any kind of improvement of fruit and its growth features. There are a lot of known genes that grant desirable traits; hence, it is no longer a barrier, also, the number of tissue specific promoters is growing steadily contributing to this transformation.

Introducing one or few genes among clonal propagated crops can help in correcting deficiencies of otherwise excellent cultivars which is a very promising futuristic approach. As reviewed by Fisk and Dandekar 1944; Schuurman and Dandekar 1994, the Fruits and Nuts species has grown significantly in the past few years by their genetic manipulation.

Modulation of genes can be done by Agrobacterium -mediated transformation which infects the plant cells by introduction of new genes. It has been regarded as a favored system for transformation of fruit trees.

However, other methods like electroporation and PEG DNA uptake in protoplasts and DNA-coated microprojectile which works upon the process of direct DNA mediated gene transfer have very limited application in fruit tree transformation.

Although any transformation using direct DNA transfer into the protoplast of fruit trees has not been reported yet but in some fruits like Citrus, Prunus, Malus, Pyrus a successful sign of regeneration from protoplast has been seen.

Particle bombardment is another process that has been used for the genetic transformation and has shown a continuous balance growth in certain cases. Genetic transformation procedure's development requires such plant cells that have the ability to regenerate plants from somatic tissues.

A lot of regeneration has been successfully achieved with use of seeds, seedlings, and several in vitro propagated commercial cultivators. However, the efficiency of such regeneration is dependent on its genotype.

Different factors including chemical, physiological, and physical factors are said to be depicting the direct influence of the developed regeneration procedures that consequently affects the frequency of regeneration in certain cultivators.

The process that has been recognized for F fruit tree species regeneration is somatic embryogenesis (formation of adventitious) or adventitious organogenesis (formation of shoots & roots). Direct and Indirect methods have been developed; however, the best of these methods has not yet been recognized, and the research is still going on in this regard.

If a bud in its earliest stage originates from more than one cell, there are chances of chimeras of transformed and non-transformed tissues occurring. In order to avoid such situations, one can pass through the callus phase or to keep strong selective pressure during transformation.

Somatic embryos have been seen in quite a few fruit species and have been proved to originate from a single cell. A walnut is the best example for studying the origin of regeneration structures since a walnut's secondary somatic embryos are formed from epidermal cell on existing somatic embryos in culture.

The process of transformation needs cells that will be useful in the regeneration process and the transformation process. There are several factors that the process of transformation depends on, they are:

For The construction of bacteria, transformational conditions, and the ability to regenerate plants from transformed tissue, the selection procedure should be proper, the development of intact transgenic plant.

Generally, the selection of transformed cells is done by herbicide resistance or by incorporating antibiotic. The most commonly used transformed cells are the neomycin phosphotransferase (NPTLL) gene that encodes to kanamycin. The assessment of the expression patterns is done by a reporter gene b-glucoronidase (GUS).

There are many factors that affect the way the introduced genes are expressed in transgenic plants. It includes the number of gene copies, the constructs, and the place of integration. However, there is apprehension in public with the products of genetic engineering. There is an ongoing field test of transgenic fruit plants and soon there will be a full-fledged marketplace for the transgenic plants in near future.

In order to verify that there is no unwanted occurrence and that the desired trait is expressed in the new plant, field testing is essential. The public is still apprehensive about the products of genetic engineering. Field tests of transgenic fruit plants are now in progress. The near future will show their place in the market.

8.11. REGENERATION AND TRANSFORMATION

8.11.1. Apple

The cultivars of Apple are one of the most successful in vitro propagated plants. The mean object for investigation is the development of the regeneration procedures for commercial cultivars. One of the most suitable source materials for transformation is the micro propagated plants because of the fact that they are genetically identical, suitable for manipulations and cultivated in the situation under control.

The morphogenesis process was affected by numerous factors, such as:

Basal nutritional medium, explant origin and maturity, cytokinin concentration, position on the medium, temperature and light.

8.11.2. Pear

There has been successful performance of the *Pints communis* having the adventitious shoots regeneration. However, bear has not been transformed but still there can be expectation of positive results soon. It could be achieved only when the regeneration is achieved and attacin E gene that induces resistance to fire blight is available.

8.11.3. Quince

In quince, there has been insertion of the reporter gene already. Under root stock of quince (*Cydonia oblonga*), the transformation was recently reported.

8.11.4. Stone Fruits

So far, there is no publication regarding the regeneration procedure in vitro propagated plants from prunus cultivars. This is the reason that the seeds and seedlings are considered as the target material for transformation.

8.11.5. Peach

The immature embryos and cotyledons were used for the development and regeneration procedure in plants. Several times, there are records of the successful reports of peach transformation (Scorza et al 1995).

The usage of seedling-derived explants is done. It has been done by introducing cytokinin biosynthesis genes (IPT) into peach. Using the biolistic approach, the plants that were stable were not isolated but the bombarding callus tissue transient gene expressions was studied.

8.11.6. Apricot

The method of getting a transgenic apricot plant is by obtaining the constructs caring coat protein of plum pox virus (PPV). In this, PCR confirm that the transformation is happening. The level of resistance that is to be evaluated is in progress.

8.11.7. Plum

Adventitious organogenesis has been used for the regeneration process to be induced in cotyledon explants of *Prunus persica*, *Prunus domestica* and *Prtmus carasus*. However, the root segments were used to regenerate the intact plants.

The regeneration procedure was developed from the leaves of bulgarian cultivar, Kjstendilska. Recently, plum was transformed using the seed hypocotyl explants.

For checking whether the plants expressing plum pox virus (PPV) is virus resistant or not, the plum was tested in the greenhouse. Also, an Austrian group has reported a successful transformation of plum with PPV.

8.11.8. Citrus

In rootstock scions, there is emphasis on the coat protein mediated resistance by the transformation research in citrus tristeza virus. In order to insert the coat protein gene of Papaya Ringspot Vims (PRSV), there has been much work done with concerns to the tropical tree fruit papaya (*Carica papaya* L.).

It was done by using the particle bombardment. There was range of reactions from the regenerated transgenic plants to PRSV.

8.11.9. Tree Nuts

With the help of seed-derived somatic embryos, walnuts can now be transformed routinely. The regeneration has been done of the lines that were transformed with the selective marker gene NPTLL and scorable marker gene gus.

The walnut plants from the transgenic method have been produced with the expression of insecticide proteins from *Bacillus thuringensis*. It is correctly showing resistance to the navel orange worm (*Amyelois transitella* Walker).

By using the walnut somatic embryo protocols, there was transformation of pecan. There is publication of the regenerated seed derived material of hazelnut via somatic embryos. However, there has been no development with respect to the transformation procedure.

8.11.10. Small Fruit

Nowadays, the transformation of strawberry cultivars and the similar fruit is now a routine. In the transgenic strawberry, the main objective is to induce resistance to mild yellow edge virus. There have been several positive results that are obtained in several laboratories.

8.12. RISK ASSESSMENT AND THE REGULATORY FRAMEWORK

8.12.1. Risk Assessment

The development and the advancement in transgenic plants have brought new techniques and responsibilities for the safe use of the plants for the betterment of humans and environment. The risk assessment aspect of the transgenic plants implies that the safety procedures are to be taken proactively rather than reactively.

The safety of the plants is dependent on scale of the growth of the plant. The measures are different in small-scale field than the large-scale fields. After the large-scale commercial production of the transgenic crops, the long-term effects of the plants could be known.

Therefore, the risk of the production of the transgenic crops is all about how it affects the health of the human beings and its effects on environment. The risk assessment can also include other issues that have emerged in time.

The things at risk are the jobs, investments, and competitiveness if the technology is not adopted. The potential benefits would be at loss. The consumer can also deny the technology at the point of sale as this comes in their rights and the issue of their choices.

This has emerged as a consumer movement in all the parts of the world and has been the point of focus, especially in Europe. Additionally, some of the international companies supports the commercialization of these new technology having the required capital and power to dominate the forming market of the transgenic world of plants.

In future, there will be just few international companies that will be in production of the recombinant plants. These companies would be helping in managing the diseases in plants by producing the industrial and agricultural products. If they do so, they will be leaving the developing countries behind with having the latest technology.

However, on a major level, the laboratories from different countries can share their advance technologies on a shred network that would ultimately help the countries to develop transgenic crops with the attest technology. The large variety of crops would cost a minor value for the companies but it would be there as a source of food and agriculture for the developing countries.

Following is the summary of possible risks that can be there for the humans and the environment with regards to the transgenic plants, these are as follows:

8.12.2. Risks for Human Health

- There can be at some stages an allergic reaction in the humans due to the formation of new allergens from the novel proteins in the transgenic organism.
- The interactions between the product of the genetic modification and other endogenous constituents of the organism could result in creation of new toxins.
- The antibiotic resistance genes were used as a marker from the food derived from genetically modified organism to gut microorganisms. The antibiotic resistant pathogens having problems have been intensified.

8.12.3. Risks for the Environment

- There can be harmful effects to the environment as there could be some species that results in the pests emerging from the gene transfer from the transgenic plant to related species by hybridization.

- Sometimes, the purpose of the transgenic plant gets distracted and it escapes its intended use and becomes an invader into the natural environment.
- The non-targeted species of the environment gets affected adversely, such as insecticide toxins can kill beneficial as well as targeted insects.
- The continuous use of the same agent on the target organism leads to the development of resistance.
- The natural biochemical cycles get adversely affected when there is interference of the transgenic plants.
- There is harmful effect of the interference of the transgenic plants on the biodiversity if there is an adaptive advantage in transgenic plants, a transgene can offer which was escaped in the area of cultivation or in wild relatives. This could have been done by cross-fertilization.

Practically, this is an important aspect if it occurs at the centers of genetic variation of cultivated plants. Moreover, there are various concerns that has been raised for the biodiversity regarding the current cultivation systems. It includes many locally adopted varieties which if gets substituted by a few new transgenics.

The examination of the above risks can help in making of the framework that can easily assess the risks for the environment. However, all the possible risks discussed above should not be considered as the concern for all the different transgenics and in different environments. But it could be used as a framework for making the process clear.

8.13. CONCLUSION

The transgenic crops in the future will be used for the improvement in agronomic traits along with the traits that involves food processing, specialty chemicals and pharmaceuticals (edible vaccines included). There is availability of the transgenic rubber tree which will be used for a variety of purposes.

Therefore, it can be assumed that there is bright and optimistic future for transgenic crops. The transgenic crops market is expected to reach heights. However, in order to achieve these goals, the sustained efforts of the industrialization and developing countries are needed.

There is a need to give attention and respond to these situations especially the public and farmers are required to respond to this changing scenario. In order to realize the benefits of these transgenic crops, the public and private sectors people have to play a significant role. This will ultimately result in the production of large number of transgenic crops in this decade.

REFERENCES

1. Boutigny, A., Dohin, N., Pornin, D. and Rolland, M., 2020. Overview and Detectability of The Genetic Modifications in Ornamental Plants. [online] Available at: <https://www.nature.com/articles/s41438-019-0232-5> [Accessed 5 June 2020].
2. Isaaa.org. n.d. Biotechnology in Ornamental Plants | ISAAA.Org. [online] Available at: <https://www.isaaa.org/resources/publications/pocketk/47/default.asp> [Accessed 5 June 2020].
3. Jhansi Rani, S. and Usha, R., 2013. Transgenic plants: Types, benefits, public concerns, and future. Journal of Pharmacy Research, [online] 6(8), pp.879-883. Available at: <https://www.sciencedirect.com/science/article/pii/S0974694313003289> [Accessed 5 June 2020].
4. Mondal, S., n.d. 8 Main Advantages of Transgenic Plant | Genetics. [online] Biology Discussion. Available at: <https://www.biologydiscussion.com/plants/transgenic-plants/8-main-advantages-of-transgenic-plant-genetics/38097> [Accessed 5 June 2020].
5. Sharma, R. and Messar, Y., n.d. Transgenics In Ornamental Crops: Creating Novelties in Economically Important Cut Flowers. [eBook] Available at: <https://www.currentscience.ac.in/Volumes/113/01/0043.pdf> [Accessed 5 June 2020].
6. Trifonova, A. and Atanassov, A., 1996. Genetic Transformation of Fruit and Nut Species. Biotechnology & Biotechnological Equipment, [online] 10(1), pp.3-10. Available at: <https://www.tandfonline.com/doi/pdf/10.1080/13102818.1996.10818873> [Accessed 5 June 2020].
7. Tsaftaris, A., Polidoros, A., Karavangeli, M., Nianiou-Obeidat, I., Madesis, P. and Goudoula, C., 2000. Transgenic Crops: Recent Developments and Prospects. Biological Resource Management Connecting Science and Policy, [online] pp.187-203. Available at: <https://www.researchgate.net/publication/200159708_Transgenic_Crops_Recent_Developments_and_Prospects> [Accessed 5 June 2020].

INDEX

N

O

P

Q

R

S

T

V

W